REGULATORY RISK: ECONOMIC PRINCIPLES AND APPLICATIONS TO NATURAL GAS PIPELINES AND OTHER INDUSTRIES

Topics in Regulatory Economics and Policy Series

Michael A. Crew, Editor
Graduate School of Management
Rutgers University
Newark, New Jersey, U.S.A.

Previously published books in the series:

REGULATORY RISK: ECONOMIC PRINCIPLES AND APPLICATIONS TO NATURAL GAS PIPELINES AND OTHER INDUSTRIES

A. Lawrence Kolbe
and
William B. Tye
The Brattle Group

Stewart C. Myers
Sloan School of Management
Massachusetts Institute of Technology
and The Brattle Group

Kluwer Academic Publishers
Boston/Dordrecht/London

Distributors for North America:
Kluwer Academic Publishers
101 Philip Drive
Assinippi Park
Norwell, Massachusetts 02061 USA

Distributors for all other countries:
Kluwer Academic Publishers Group
Distribution Centre
Post Office Box 322
3300 AH Dordrecht, THE NETHERLANDS

Library of Congress Cataloging-in-Publication Data

Kolbe, A. Lawrence.
    Regulatory risk : economic principles and applications to natural
gas pipelines and other industries / A. Lawrence Kolbe and William
B. Tye, Steward C. Myers.
       p. cm. -- (Topics in regulatory economics and policy series ;
14)
    Includes bibliographical references and index.
    ISBN 0-7923-9330-9
    1. Gas industry--Government policy--United States. 2. Gas
pipelines--Government policy--United States. 3. Gas companies-
-Rates--Government policy--United States. 4. Public utilities-
-Rates--Law and legislation--United States. 5. Public utilities-
-United States--Rate of return. 6. Risk. I. Tye, W. B. (William
B.) II. Myers, Steward C. III. Title. IV. Series: Topics in
regulatory economics and policy ; 14.
HD9581.U5K64 1993   93-12516
388.5'6'0973--dc20   CIP

Printed on acid-free paper.

Printed in the United States of America

Table of Contents

Acknowledgements

In connection with the development of the basic theory, the authors wish to thank Alan Buchmann, Michael Dawson, Dick Disbrow, Ernest Ellingson, Victor Goldberg, Skip Horvath, Gibson Lanier, Nancy Lo, John Meyer, Jon Noland, Richard Pierce, Richard Rapp, Charles Stalon, John Strong, Judge Stephen Williams, and the editors of the *Yale Journal on Regulation* for helpful comments.

With regard to the application of the theory to natural gas pipelines, the authors thank Skip Horvath of the Interstate Natural Gas Association of America (INGAA) for his many hours helping us understand the regulatory and operational details of the gas pipeline industry. We also thank former FERC Commissioners George Hall and Charles Stalon for similar assistance and for their comments on earlier drafts. We also have discussed these issues with many other people, both within and outside the pipeline industry, and we hereby thank them all.

We also thank Jamie Simler and Paul Hoffman of INGAA for helping us gather data on the industry. Within *The Brattle Group* Carlos Lapuerta reviewed many documents, analyzed the data, and critiqued our logic. Helen Ferris provided excellent word processing and editorial assistance. Faye Mazo assisted with preparation of the figures, and other staff pitched in as the occasion required. Mary Ann Buescher edited the manuscript and prepared the index. We thank all of them for their efforts.

Of course, the conclusions and content remain the sole responsibility of the authors.

Copyright Acknowledgements

Grateful acknowledgement is given for permission to reprint extracts of the following:

- Reprinted with permission from *Yale Journal on Regulation*, "The *Duquesne* Opinion: How Much 'Hope' is There for Investors in Regulated Firms?" by A. Lawrence Kolbe and William B. Tye, c 1991, Yale Law School.

- Reprinted with permission from *Research in Law and Economics*, "The Fair Allowed Rate of Return with Regulatory Risk," by A. Lawrence Kolbe and William B. Tye, c 1992, JAI Press, Inc.

- Much of Chapters 4 and 6 through 9 is from *Risk of the Interstate Natural Gas Pipeline Industry*, by A. Lawrence Kolbe, William B. Tye, and Stewart C. Myers (Washington, D.C.: Interstate Natural Gas Association of America, 1991).

REGULATORY RISK:
ECONOMIC PRINCIPLES AND
APPLICATIONS TO
NATURAL GAS PIPELINES AND
OTHER INDUSTRIES

Chapter 1
Introduction

It is perfectly obvious that the concept of "risk" has
taken on wholly new dimensions in recent decades and
is today being reflected on in ways that would have been
almost inconceivable even a few years ago. The older
idea, that risk is essentially a wager, which individuals
make in the hope of gaining something significant, substantial, has almost disappeared from common parlance.
Risk today is conceived principally as danger, great
danger, which is not run out of choice and cannot be
said to derive from the hazards associated with certain
kinds of advanced technology. Increasingly, in the
United States perhaps even more than elsewhere, the tendency is to be preoccupied with what are perceived to be
the unnecessary risks of contemporary life, against which
the individual needs to be protected by the national government (Graubard 1990).

This book is the result of an intellectual odyssey going back more than
10 years. For us, the discovery that there was a fundamental flaw in the
entire method for evaluating the magnitude of asymmetric risks and
compensating investors in regulated industries for bearing those risks
came in the early 1980s. At that time, we were working on the seminal
case involving oil pipeline regulation before the Federal Energy Regulatory Commission (FERC).

The particular issue that initiated our inquiry was the appropriate depreciation schedule for oil pipelines. No one knew what the economic lifetime of a pipeline would be. This uncertainty presented a problem,
because a pipeline whose economic lifetime was prematurely truncated
would expose the investor to the risk of a 'stranded asset' whose depreciation might not be recovered from ratepayers. (AT&T wrote off billions
of dollars of such investments in obsolete equipment upon divestiture.)
If the oil reserves outlasted the regulatory depreciation life, on the other
hand, the company would possess a 'zero rate base' asset with no incentive to use it despite its obvious continued value to customers. It was
clear that changes in regulatory depreciation policy could have a major
impact on the risk of holding stranded assets. Regulatory depreciation
policy was hardly a matter of indifference to investors.

Yet when we examined the traditional methods for estimating the cost of capital used by experts before regulatory commissions, nothing happened to the estimate under most common methods when regulators change the depreciation formula! This was a real problem, because the experts usually argue that investors have been fairly treated if the regulatory commission sets the revenue requirement so that the allowed rate of return on assets in the rate base equals the cost of capital. Something was amiss, it was clear, in the fundamental methodology being used to solve the central problem in rate of return regulation.

Other examples of blatant inconsistencies between the methods for establishing the allowed rate of return and the risks inherent in the regulatory scheme also came to our attention. After tens of thousands of pages of testimony in Phase I of the Trans Alaska Pipeline System (TAPS) rate case, an Administrative Law Judge at the FERC deemed TAPS to be a low risk pipeline deserving a rate of return on rate base even below the then prevailing prime interest rate. The Commission then announced that it would proceed to Phase II, where it would consider whether to disallow billions of dollars of construction expenses because of the cost overrun! Clearly, the method for setting the rate of return in Phase I had taken no account of extreme regulatory risk in Phase II.

The authors spent the next few years giving speeches that no one understood and writing papers demonstrating the fundamental error. The papers went unpublished and the speeches went unheard.

We were voices crying in the wilderness until Michael McBride of the law firm of LeBoeuf, Lamb, et al., sent us the briefs of the parties in *Duquesne Light Co.* v. *Barasch*, 109 S. Ct. 609 (1989), a case before the U.S. Supreme Court involving a switch in regulatory methodology with adverse consequences for investors. Moreover, the case incorporated unusual facts: the switch was explicit (incorporated in a new law), retroactive (passed during the last week of hearings before the public utility commission), and no allegations of imprudence were made (the investment was unused and unuseful, but everybody agreed it was prudent). The result of the retroactive switch was to disallow the recovery from ratepayers of millions of dollars of prudently incurred investment.

Reading the briefs to the Supreme Court, one could see that the case involved the very economic issues we had struggled with over the last ten years, albeit couched in the terms of constitutional law -- 'takings,' 'due process,' etc. A financial analyst who understands the issue of regulatory risk could only say, "Put me in, Coach."

The Supreme Court, interestingly, put aside all the constitutional arguments of the parties and, without benefit of briefing, undertook an economic analysis in the opinion. And the economics were not bad. It certainly was a lot more cogently reasoned than it would have been if the misconceptions of the traditional regulatory practice had found their way into the briefs and oral arguments.

Normally, it takes twenty to thirty years for a new economic idea to (1) start out as a minority view, (2) become widely accepted among economists, (3) be reflected in commission orders and lower court opinions, and (4) finally be enshrined in a Supreme Court opinion. (By this time it is often no longer the conventional wisdom as an emerging skeptical literature is undercutting the consensus view that started the process.) In the case of regulatory risk, the Supreme Court 'skipped the middlemen' and embodied the new idea directly, saving the country twenty or thirty years of arguing.

Remarkably, the Supreme Court got the basic economic ideas right, while making a crucial presumption of fact based on evidence not in the record and getting some of the calculations wrong. While most observers in the regulated industries believed that the Supreme Court had punted on the biggest regulatory crisis of the century, we felt that the Court had come to many (but not all) of the same conclusions we had.

More specifically, the Court ruled in *Duquesne* that:

1. 'Regulatory risk' is a special class of risk that must be recognized by regulators when setting the allowed rate of return.

2. A potential future disallowance of a return *of* and/or a return *on* unused and unuseful investments is a class of regulatory risk.

3.   The Court *presumed* that the additional regulatory risk experienced by investors under the ratemaking standard in use prior to the switch in ratemaking methodology was offset by a higher allowed rate of return.

4.   "Serious constitutional questions" would be raised if regulators opportunistically switched ratemaking methodologies to shift exposure to downside risks to investors after adverse outcomes were observed.

5.   The switch in regulatory methodology which occurred in *Duquesne* did not present a question of "constitutional magnitude" because the "end result" of the switch in methodology was "slight" and within the range of the plausible end results of the prudent original cost methodology approved in the *Hope Natural Gas* opinion.

The Court's conclusions 1 and 2 were exactly what we had been unsuccessfully arguing for years to no avail. The third ruling was indeed a presumption, because it was unsupported by facts in the record and is completely contrary to typical practice. In their defense, however, it is easy to see that the Court would not believe without further proof that generations of experts and regulators could be so seriously wrong for so long about a fundamental proposition that seemed so obvious.

The fourth conclusion is one of law upon which we do not now comment, other than to observe that the risk of opportunistic switches in ratemaking methodology is yet another example of regulatory risk. Lastly, we find that the Court's method for estimating the magnitude of switches does not conform to generally accepted principles of financial economics.

People could ignore us, but they could not ignore the Supreme Court. In its first decision on ratemaking principles in 45 years, we believe that the Supreme Court was saying something new and different.

We rewrote our previously unpublished papers, but this time integrated them with a discussion of the *Duquesne* opinion (see Chapter 2). Economics and the law were coming together, and the *Yale Journal on Regulation* agreed to publish our paper. People started to understand what we were saying. Indeed, some observers were now saying that the new principles of economic regulation called for by explicit recognition of regulatory risk were so obvious that they hardly needed repeating. Our views, previously rejected on the grounds they were wrong, were now being rejected because the results were too obvious to publish! A significant minority, including a large percentage of the members of the Economic Committee of the Edison Electric Institute (a trade association for electric utilities), continued to say we were wrong, however.

A real breakthrough came when we were retained by the Massachusetts gas utilities to analyze the issue of cost recovery for environmental cleanup liabilities from 'blue dirt,' i.e., the solid waste from coal gasification during the 'gaslight era' prior to the arrival of natural gas to the Northeast. The total money at stake was more than half of the net worth of the firms involved.

Prior to the introduction of the economic principles of regulatory risk and the legal principles of *Duquesne*, the two positions might well be perceived by regulators as "'Tis" and "'Tain't." The regulated utility often loses in this case.

Using the economic principles of *Duquesne*, it became clear that the risk of disallowance of the cost recovery for environmental cleanup expenses is no different in principle than the risk of disallowance of used and useful investments addressed in *Duquesne*. But was the risk of such disallowances previously accounted for in the allowed rate of return, as the Supreme Court *presumed* in *Duquesne*? The answer is "no" under all of the accepted methodologies for computing the cost of capital (see Chapter 3 for discussion). If so, then investors had not been compensated historically for the risk of the environmental cleanup costs, and the ratepayers should properly pay the costs.

The logic was compelling enough that the case settled on terms deemed reasonable to the utilities. Terms of the settlement were:

1.   No prudence review;

2.   Amortization of *all* costs over a seven-year period, without a carrying charge, up to 5 percent of previous year's revenues (i.e., "return of but not on" outlays);

3.   A carrying charge on amounts beyond 5 percent until they can be amortized;

4.   No request by the companies for higher allowed rate of return based on this treatment; and

5.   Companies can keep profits on future sales of the land up to amount of the lost carrying charges.

Shortly thereafter we were retained by the Interstate Natural Gas Association of America (INGAA) to do a study of risk for interstate natural gas pipelines. The CEOs of the firms were convinced that the industry was highly risky, despite regulatory folklore. The industry had lost billions in unrecoverable gas acquisition costs in the 1980s collapse of energy prices. Yet allowed rates of return in federal regulatory proceedings were going down, not up. Furthermore, gas acquisition costs were historically recovered with a Purchased Gas Adjustment (PGA) mechanism that allowed no profit on gas itself.

Now one can argue whether an additional component for regulatory risk is included in a positive allowed rate of return on assets in the rate base. But positive compensation for assuming regulatory risk of unrecoverable gas acquisition costs cannot be included in the number zero (the allowed profit margin on acquired gas). Gas industry senior executives were convinced that there was something dreadfully wrong with the system, but no one was quite sure exactly what assumption in the regulatory process was wrong.

The distinction between the fair allowed rate of return and the cost of capital which was developed in the *Yale Journal on Regulation* article explained much of what was going on. No matter how good a job financial experts and regulators did in computing the cost of capital, the resulting number was seriously biased as a measure of the proper fair allowed rate of return, because it did not address the basic asymmetry of the cost recovery policy for gas acquisition. The resulting study of the interstate gas pipeline industry is our most comprehensive application yet of the phenomenon of regulatory risk.

Having developed the theory, the task in this book is to flesh out the practical regulatory mechanisms to put the theory into practice. This book seeks to accomplish this implementation by first providing a coherent explanation of the basic theory. Chapter 2 may thus be skipped by those already familiar with the outline of the arguments in the published *Yale Journal on Regulation* piece. Chapter 3 replies to our critics, who curiously object to the theory chiefly on two absolutely inconsistent grounds. This, too, may be skipped by those who are familiar with our article in *Research in Law and Economics* or those who accept the results of Chapter 2 and are chiefly interested in the application to pipelines and other regulated industries.

Chapter 4 begins our application of these principles to the interstate gas pipeline industry by first identifying the sources of regulatory and other asymmetric risk. Risks are characterized into those that affect the cost of capital and those that do not. Of the later category it is demonstrated that asymmetry is the key property that must be recognized. Appendix A provides more details of this classification scheme for risk. Chapter 5 provides the conclusions for the application of the principles to gas pipelines.

The path leading to those conclusions starts with Chapter 6, which identifies financial and other trends affecting the industry. Appendix B supplements Chapter 6 by providing a detailed history of the industry. Chapter 7 identifies two views of what these trends mean for risk in the industry. Chapter 8 combines the theory and data from the previous chapters to analyze the risk of the pipeline industry. Chapter 9 uses these conclusions to evaluate proposed regulatory schemes for the industry in the future.

Chapter 2
The Theory of Regulatory Risk

The meaning of "fairness" in business transactions is most clearly definable when referring to a moral obligation, which may also be a legal obligation, to avoid deception and to live up to previous commitments, expressed or implied. If judged by this test alone, any rule of rate making would be fair to investors, whatever its merits or demerits on other grounds, if it conforms to the terms, on the faith of which the investment was originally made -- fair no matter how onerous or how profitable these terms may prove to be in the light of hindsight (Bonbright 1961, 127).

INTRODUCTION

High costs for new electric power plants have led to a series of regulatory and legislative decisions that may retroactively rewrite the rules that utility investors relied upon when they supplied capital for these projects. In *Duquesne Light Co.* v. *Barasch*, 109 S.Ct. 609 (1989), a case involving the recovery of capital invested in constructing nuclear power plants that were ultimately never completed, the Supreme Court upheld one such change.

The effect of this decision is to permit state regulators to shift to investors losses from utility assets that are never used and never shown to be useful. The Court made five key rulings on economic and financial matters which are repeated here for the convenience of the reader:

1.   'Regulatory risk' is a special class of risk that must be recognized by regulators when setting the allowed rate of return.

2.   A potential future disallowance of a return *of* and/or a return *on* unused and unuseful investments is a class of regulatory risk.

3.   The Court *presumed* that the additional regulatory risk experienced by investors under the rate-making standard in use prior to the switch in

ratemaking methodology was offset by a higher
allowed rate of return.

4.   "Serious constitutional questions" would be
raised if regulators opportunistically switched
ratemaking methodologies to shift exposure to
downside risks to investors after adverse outcomes were observed.

5.   The switch in regulatory methodology which
occurred in *Duquesne* did not present a question
of "constitutional magnitude" because the "end
result" of the switch in methodology was
"slight" and within the range of the plausible
end results of the prudent original cost methodology approved in the *Hope Natural Gas* opinion.[1]

The thesis of this chapter is that it may prove difficult in practice to
provide the compensation for regulatory risks that the first ruling requires, especially given the fifth ruling that "slight" policy changes are
permitted. Specifically, in support of its fifth key ruling, the opinion
cites two definitions of slight, but both recognize only a fraction of the
actual loss. If these definitions were taken as a required precedent, even
quite large takings might be dismissed as not of constitutional stature.
Duquesne may further restrain the start of new investments under the
traditional form of ownership in regulated industries.

In the regulatory environment after *Duquesne*, investors are exposed to
substantial risks from very large cost disallowances without equivalent
opportunities for gain. This asymmetry in regulatory outcomes (unless
otherwise corrected) requires a rate of return in excess of the cost of
capital, defined as the expected rate of return in capital markets on alternative investments of equivalent risk. These findings are fully in the
spirit of the Supreme Court's express intention in *Duquesne* to reaffirm
the teachings of *Federal Power Commission* v. *Hope Natural Gas* (here-

---

[1]   320 U.S. 591 (1944).

inafter, '*Hope*'). The *Hope* opinion requires that the "return to the equity owner should be commensurate with return on investments in other enterprises having corresponding risks."[2]

This chapter examines the previously noted rulings in *Duquesne* and their economic implications for utility investors. In doing so, we employ principles of financial analysis to reevaluate what were thought to have been settled economic and constitutional questions on the mechanics of public utility regulation. As shown below, the relevant economic concepts require that in the presence of material asymmetric risk, the allowed rate of return be in excess of the cost of capital or that some equivalent "insurance premium" be paid to investors. This conclusion implies the need for substantial change in public utility regulation.

The next three sections describe our path to this conclusion. The facts and opinion of the *Duquesne* case represent a particularly interesting example of the economic issues. We also find that the Court's opinion in many respects demonstrates a more sound understanding of the relevant concepts than is embodied in current practice and much expert opinion. The first section therefore introduces the subject with a fairly detailed discussion of the circumstances of the case before the Court and the logic of the opinion. Because a clear understanding of the issues requires a sound understanding of the relevant economic principles, the second section develops the essential concepts of financial economics. The third section applies these principles to the issues identified in the first section. A concluding section looks at the implications for regulation and the utility industry.

Duquesne

At issue in *Duquesne* was a state statute that was used to deny recovery of capital invested in several nuclear plants by use of a new 'used and useful' test.[3] The particular legal question facing the Court was whether

---

[2]   320 U.S. 603.

[3]   In this context, 'used and useful' means the plants are completed and used to provide service to customers (Kahn 1985; Pierce 1983).

the statute violated a prior 'regulatory contract'[4] and thereby represented an unconstitutional taking of property. The Court's opinion held that under the 'end result' test,[5] the new rules did not reach the stature to constitute a taking in this case. The decision also left considerable room for future regulatory discretion.[6]

Hope and the Traditional Economic Paradigm of Rate Regulation

Rate of return and rate base methodology for the regulated firm is probably the oldest issue in rate regulation.[7] Without some idea of whether the rate of profit as a percent of asset value is high or low, the regulator has little idea of whether the regulated firm's market power (if any) is constrained by regulation.

Between 1898 and 1944, the Court handed down a number of decisions on whether the original cost or the fair value of an investment was the

---

[4]   For some time it has been recognized that from an economic point of view, regulation may be thought of as an implicit contract governing long-term relationships between buyers and sellers (Goldberg 1976). Fair regulation has generally proscribed the use of 'retroactive ratemaking' to revise the observed results of a previously agreed upon ratemaking approach.

[5]   The end result test refers to the Supreme Court's statement in *Hope*:

> Under the statutory standard of "just and reasonable" it is the result reached not the method employed which is controlling [citations omitted]. It is not theory but the impact of the rate order which counts. If the total effect of the rate order cannot be said to be unjust and unreasonable, judicial inquiry under the Act is at an end. The fact that the method employed to reach that result may contain infirmities is not then important. 320 U.S. 602.

[6]   The *Duquesne* decision states that "the Constitution within broad limits leaves the States free to decide what rate-setting methodology best meets their needs in balancing the interests of the utility and the public." 109 S.Ct. at 620.

[7]   Alfred E. Kahn (1970, 20) states that "price regulation is the heart of public utility regulation" and discusses how that proceeds from a determination of the appropriate rate base and rate of return. He discusses the history of the debate at pp. 35-54.

better rate base.[8] The Court seemingly resolved the matter in *Hope*. According to *Hope*, what mattered was the "overall effect," "total effect," or more commonly, the end result.[9] The use of an original cost less depreciation rate base was found to satisfy this test. Following *Hope*, the rate base became essentially (but not always) a known quantity, prudently invested original cost less depreciation.[10] The focus of debate then became the determination of the appropriate rate of return.[11] If investors were to be compensated for differences in risk and inflation, the adjustment would be to the rate of return, not the rate base.

Given a fixed investment base, pricing of the services of regulated industries was historically done on a bottom-up 'cost-of-service' basis, whereby the firm's expenses were added to an allowance for interest, profit, and income taxes on the prudent original cost investment base. The *Hope* court endorsed its earlier definition of the profit component in

---

[8]   See below at footnote 35 for a discussion of this important debate over the fundamental concepts of ratemaking, and Kahn (1970, 38-41).

[9]   *Hope*, 320 U.S. 602.

[10]   The prudent original cost test had been advocated for some time by Justice Brandeis. See *State of Missouri ex rel. Southwestern Bell Telephone Co.* v. *Public Service Commission*, 262 U.S. 276, 292 (1923) (Brandeis, J. concurring). Some states kept fair value rate bases by statute, but a common belief is that regulatory commissions in these states tried to set the rate of return on those fair value rate bases so the end result in each year was the same that would have been reached with an original cost rate base and a rate of return appropriate for that. We cannot verify or refute this view, but we assume it is correct for purposes of this paper, at least for electric utilities. Recently there have been attempts at non-original-cost rate bases with different end results year by year. Perhaps the leading example is the Federal Energy Regulatory Commission's "trended original cost" rate base for oil pipelines. See Stewart C. Myers, A. Lawrence Kolbe, and William B. Tye (1985) for discussion of the economics of original cost versus trended original cost. All of the fundamental conclusions of the present paper remain true under trended original cost, but some of the implementation details would change.

[11]   109 S. Ct. at 617.

Bluefield Waterworks & Improvement Co. v. *Public Service Commission* (hereinafter, '*Bluefield*').[12] *Bluefield* had held that a regulated firm

> is entitled to such rates as will permit it to earn a return on the value of the property which it employs . . . equal to that generally being made . . . on investments in other business undertakings which are attended by corresponding risks and uncertainties.[13]

Hope itself required that the investor's return be commensurate with returns on equally risky investments in other enterprises.[14] The modern economic view is that this condition is satisfied when investors expect to earn the cost of capital,[15] defined as "the expected rate of return in capital markets [e.g., the New York Stock Exchange] on alternative investments of comparable risk."[16]

Thus, in the traditional economic paradigm of rate regulation, customers expect to pay (and investors to earn) a just and reasonable rate of return on the actual costs of investments, which are recorded in the firm's rate base. These costs might be higher or lower than originally expected, but all are put in the rate base as long as they were prudently incurred. Hence the term, pure prudent investment rule.

---

[12]   262 U.S. 678 (1923).

[13]   *Id.* at 692.

[14]   *Hope*, 320 U.S. at 603.

[15]   See Stewart C. Myers (1972) for the modern economic definition of a 'just and reasonable' rate of return.

[16]   See, for example, Richard A. Brealey and Stewart C. Myers (1988, Chapter 7), and A. Lawrence Kolbe and James A. Read, Jr., with George R. Hall (1984, Chapter 2).

Background to the Duquesne Decision

In 1967, Duquesne Light Company and four other utilities joined a venture (CAPCO[17]) to construct seven nuclear generating units. In 1980, four of the plants were canceled because of the economic and political impacts of the Arab oil embargo, the accident at Three Mile Island, and other intervening events.[18]

The Pennsylvania Public Utilities Commission (PUC) approved the amortization of the investment in canceled plants over a 10-year period through rate increases in 1983. However, about a month before the close of the rate case in 1982, the Pennsylvania legislature enacted a law that precluded inclusion of costs of construction of facilities in rate bases, prior to the time such facilities were "used and useful in service to the public."[19]

A consumer group sued Duquesne Light and the PUC under the new law. In the ensuing litigation, the PUC maintained that its decision to permit amortization of the aborted plants (and thus permit a return *of*

---

[17]   'CAPCO' stands for Central Area Power Coordination Group. Duquesne Light and Pennsylvania Power Company participated with three other utilities.

[18]   As a result of the Arab oil embargo, the price of energy increased, which drove up the price of electricity and drove down the rate of growth of electric demand and hence the need for the projects contemplated by CAPCO. The accident at Three Mile Island, of course, raised questions about the future of nuclear power. For a discussion of the change in economic environment that made certain investments unused and unuseful, albeit prudently made, see *Jersey Central Power and Light Co.* v. *Federal Energy Regulatory Commission*, 810 F.2d 1168 (D.C. Cir., 1987).

[19]   The statute states:

>   the cost of construction or expansion of a facility undertaken by a public utility producing . . . electricity shall not be made a part of the rate base nor otherwise included in the rates charged by the electric utility until such time as the facility is used and useful in service to the public. Pa. Cons. Stat. Ann., Tit. 66, §1315 (Purdon Supp. 1988), as cited 109 S.Ct. at 613.

investment) without inclusion in the rate base[20] (and thus not provide a return *on* investment), complied with the new Act. The Pennsylvania Supreme Court disagreed and held that the statute prohibited both a return of and a return on invested capital. In so doing, that court held that the law did not violate the 'takings' clauses of the Fifth and Fourteenth Amendments.[21]

Critical Parts of the Opinion

The U.S. Supreme Court affirmed the Pennsylvania Supreme Court decision. The precise language of the *Duquesne* opinion must be read very carefully if we are to understand its meaning. Because terms may have different meanings for judges and financial economists, and because it will be convenient to refer to these rulings throughout the remainder of this chapter we quote at length from the opinion.

---

[20]   Approved investor cash flows = [(the allowed rate of return) x (the undepreciated cost of the assets employed, called the rate base)] + an annual depreciation charge.

[21]   Regulatory standards for ratemaking have ultimately been derived by the courts from two provisions in the U.S. Constitution:

   (1)   [N]or shall private property be taken for public use, without just compensation. *U.S. Constitution*, Amend. V, cl. 4.

   (2)   [N]or shall any State deprive any person of . . . property, without due process of law. . . . *U.S. Constitution*, Amend. XIV, cl. 3.

The Supreme Court asserted its jurisdiction over the reasonableness of railroad rates in *Chicago, Milwaukee & St. P. Ry.* v. *Minnesota*, 134 U.S. 418 (1890) and reaffirmed its responsibility under the Fourteenth Amendment in *Smyth* v. *Ames*, 169 U.S. 466 (1898). The methodology adopted in *Smyth* v. *Ames* was overturned by *Hope* 320 U.S. at 591 in 1944, but not the requirement that rates be in accord with the Fifth and Fourteenth Amendments.

The Court held that a state utility regulatory scheme "does not 'take' property simply because it disallows recovery of capital investments that are not 'used and useful in service to the public.'"[22]

The Court then noted that according to the initial evaluation of the utilities' actions, "'the CAPCO decisions in regard to the [canceled plants] at every stage to their cancellation, were reasonable and prudent.'"[23] Whether a rate would pass muster as "just" and "reasonable," however, would "depend to some extent on what is a fair rate of return given the risks under a particular rate-setting system, and on the amount of capital upon which the investors are entitled to earn that return."[24]

The modified regulatory regime in effect before the new statute raised the risks of investing in the utility. The Court assumed, however, that the regime compensated for the increased risk by revising the rate of return.

> Pennsylvania has modified the [pure prudent investment[25]] system in several instances, however, when prudent investments will never be used and useful. For such occurrences, it has allowed amortization of the capital lost, but does not allow the utility to earn a return on their investment. . . . The loss to utilities from prudent but ultimately unsuccessful investments under such a system is greater than under a pure prudent investment rule, but less than under a fair value approach. Pennsylvania's modification slightly increases the overall risk of investments in utilities over the pure prudent investment

---

[22]   *Duquesne*, 109 S.Ct. at 612.

[23]   *Id.* at 613.

[24]   *Id.* at 617.

[25]   We identify below the four regulatory systems that the Court distinguishes.

> rule. *Presumably the PUC adjusts the risk premium element of the rate of return on equity accordingly.*[26]

A theoretical inconsistency in a particular rate, in and of itself, was held not to be subject to Constitutional question if the rate was otherwise reasonable. "Inconsistencies in one aspect of the methodology have no constitutional effect on the utility's property *if they are compensated by countervailing factors in some other aspect.*"[27]

While the move from one regulatory regime to another could run afoul of the Constitution, the Court stated that Pennsylvania's action did not.

> *The risks a utility faces are in large part defined by the rate methodology* because utilities are virtually always public monopolies dealing in an essential service, and so relatively immune to the usual market risks. Consequently, a State's decision to arbitrarily switch back and forth between methodologies in a way which required investors to bear the risk of bad investments at some times while denying them the benefit of good investments at others would raise serious constitutional questions. But the instant case does not present this question.[28]

According to the Court's calculation, the impact of the switch upon investors and the utilities was not significant.

> In fact the overall effect is well within the bounds of *Hope*, even with total exclusion of CAPCO costs.

---

[26]   *Id.* at 617-18 [emphasis added].

[27]   *Id.* at 619 [emphasis added].

[28]   *Id.* at 619 [emphasis added].

> Duquesne[29] was authorized to earn a 16.14 percent re-
> turn on common equity and an 11.64 percent overall
> return on a rate base of nearly $1.8 billion [citation
> omitted]. Its $35 million investment in the canceled
> plants comprises roughly 1.9 percent of its total base.
> The denial of plant amortization will reduce its annual
> allowance by 0.4 percent.[30]

The Court thus held that the losses suffered were sufficiently "slight" to avoid Constitutional questions under the takings doctrine.

> Given these numbers, it appears that the PUC would
> have acted within the constitutional range of reasonable-
> ness if it had allowed amortization of the CAPCO costs
> but set a lower rate of return on equity with the result
> that Duquesne and Penn Power received the same reve-
> nue they will under the instant orders on remand. The
> overall impact of the rate orders, then, is not constitu-
> tionally objectionable. No argument has been made that
> these *slightly reduced rates* jeopardize the financial
> integrity of the companies, either by leaving them insuf-
> ficient operating capital or by impeding their ability to
> raise future capital. Nor has it been demonstrated that
> these rates are inadequate to compensate current equity
> holders for the risk associated with their investments
> under a modified prudent investment scheme [footnote
> omitted] . . . An otherwise reasonable rate is not sub-
> ject to constitutional attack by questioning the theoretical
> consistency of the method that produced it.[31]

---

[29] Pennsylvania Power Company was also an appellant, and its stated losses were of similar magnitude.

[30] *Id.* at 618.

[31] *Id.* [emphasis added].

Regulatory Risk in Four Ratemaking Regimes

In reaching these conclusions, the Court distinguished four regulatory ratemaking regimes. Under the (1) *pure prudent investment* standard, all prudent investments go into the rate base regardless of whether they are used or useful.[32] The investor thus would receive both a return *of* and a return *on* capital for the canceled plants.

Under the (2) *modified prudent investment* standard, unused and unuseful investments were amortized but did not go into the rate base. The investor would have received a return *of* but not *on* the unused and unuseful investment under this second standard. According to the Court, this was the standard in effect before the new law was passed.[33] Investors' expectations under this second standard were not fulfilled, but not because the investments turned out to be unused and unuseful. They were unfulfilled because (2) the modified prudent investment test was revoked by the state legislature at the last minute and replaced by another test.

---

[32]   The Court stated this approach in the following passage:

> Justice Brandeis had advocated an alternative approach as the constitutional minimum, what has become known as the "prudent investment" or "historical cost" rule. He accepted the *Smyth* v. *Ames* eminent domain analogy, but concluded that what was "taken" by public utility regulation is not specific physical assets that are to be individually valued, but the capital prudently devoted to the public utility enterprise by the utilities' owners. . . . Under the prudent investment rule, the utility is compensated for all prudent investments at their actual cost when made (their "historical" cost), irrespective of whether individual investments are deemed necessary or beneficial in hindsight. The utilities incur fewer risks, but are limited to a standard rate of return on the actual amount of money reasonably invested. (Citing *Missouri* v. *Public Service Commission* 262 U.S. 276, 291 (1923); footnote omitted.)

[33]   The utilities asserted that they had historically been on the pure prudent investment standard, but did not contest the disallowance of a return on investment, leading to the Court's finding that the utilities were constructively acting under the modified prudent investment rule. 109 S.Ct. at 617-618. An explanation consistent with the utilities' view and their failure to protest is that they felt the modified prudent investment treatment was the best they could hope to attain in the current political climate and that protesting it might increase the chance of even more severe treatment.

This latter test is called (3) the *used and useful* test.[34] Under this standard the investor received neither a return of nor a return on the canceled plants.

The Court discussed a fourth test, which it labeled (4) fair value[35] and defined as follows:

> In theory, the *Smyth v. Ames* fair value standard mimics the operation of the competitive market. To the extent utilities' investments in plants are good ones (because their benefits exceed their costs) they are rewarded with an opportunity to earn an "above-cost" return, that is, a fair return on the current "market value" of the plant. To the extent utilities' investments turn out to be bad ones (such as plants that are canceled and so never used and useful to the public), the utilities suffer because the

---

[34] Under the used and useful test, only assets which are "used and useful in service to the public" will generate a return on and return of investment in the allowed revenue requirement, regardless of whether or not the investment was prudent. In the case of CAPCO it was ruled that the investments in question were unrecoverable in rates because the plants were abandoned, even though it was ruled that the decision to incur the costs was prudent. For a discussion of the move toward this test, see, for example, James J. Hoecker (1987, 303).

[35] We use the Court's characterization for the purposes of this chapter, but there have been many definitions of the term fair value. Indeed, this was responsible for much of the problem of implementing the standard. The fair value standard emerged from the Supreme Court's decision in *Smyth v. Ames*, 169 U.S. 466 (1898), where the Court listed a number of factors which could be used to ascertain that value. A. Kahn (1970, 38) states the consensus opinion that "*Smyth v. Ames* [was] the bane of public utility regulation for the next 50 years, embroiling commissions and courts in endless controversies about the measurement of fair value." The Court in Duquesne states that the fair value standard ". . . suffered from practical difficulties which ultimately led to its abandonment as a constitutional requirement." 109 S.Ct. at 616. As mentioned in footnote 32, above, the *Hope* test effectively replaced the fair value rate base with prudent original cost, and the debate among the interested parties thereafter focused on the rate of return to be applied to that rate base. Under the net original cost system that came into wide use in the years following *Hope*, the rate base until recently was subject to much less controversy than under *Smyth v. Ames*.

investments have no fair value and so justify no return.[36]

The Court ranked the four rate base methodologies according to the risks
to which investors were exposed. Pure prudent investment insulated
investors from both upward and downward future revaluations of the
investment base and thus incurred "fewer risks."[37] The modified prudent investment standard exposed investors to some loss. Under this
test, they could be denied a rate of return on unused and unuseful investments but could still recover the investment itself. The Court ranked this
standard as "slightly" more risky than the first test.[38]

Under the used and useful test, investors could lose everything but had
no prospect of capital gains if the investment was particularly valuable.
The fourth test, fair value, also presented substantial risk since asset
revaluations could go up or down depending on subsequent economic
developments.

Inasmuch as investor risk increased with movements from pure prudent
investment to modified prudent investment to the used and useful test, the
allowed rate of return would have to increase accordingly. Fair value
was ranked higher in terms of risk than pure prudent investment (because
investors were exposed to capital gains and losses). The Court noted that
potential losses were greater under fair value than under modified prudent investment, but did not explicitly rank the fair value test overall for
risk.

The Duquesne utility's investment decision and the PUC's initial determination of rate of return were made under the modified prudent investment rate base methodology, which the Court found to be in effect in

---

[36]   *Hope* at 616.

[37]   *Id.* at 616.

[38]   *Id.* at 618.

1967.[39] The Pennsylvania legislature, however, enacted the much riskier used and useful rate base test some 15 years later. Application of that test, the issue decided by the Court in *Duquesne*, of course, bears implications for investors and for utilities' ability to raise capital.

ECONOMIC GROUNDWORK

This part of the chapter sets out the economic terms and arguments upon which our analysis relies. We define four terms: (1) promised (or target) rate of return; (2) expected rate of return; (3) realized rate of return; and (4) regulatory risk. To define regulatory risk, which comes in two varieties, we contrast the traditional economic paradigm of rate regulation with the regulatory rules endorsed by *Duquesne*. We also examine whether regulators typically allow an expected or a promised rate of return on equity.

Expected Versus Promised Returns in Competitive Capital Markets

The concept of a promised rate of return can be illustrated by corporate bonds. Consider a 'junk bond,' i.e., one with an appreciable risk of default. The borrower promises investors a relatively high return, say 13.5 percent for this example; but the return investors expect is lower, because they know the borrower may default, i.e., fail to redeem the bond for the full promised amount. The situation is illustrated in Figure 2-1. The realized rate of return on the junk bond, i.e., the actual rate of return that is in fact observed at some future date, can never exceed 13.5 percent but may go substantially below it.

To determine the expected rate of return on the bond, we must know more about the likelihood of possible outcomes. For example, suppose the bond is offered at a price of $1,000, that it will be redeemed in one year when both principal and interest are paid, and that lenders agreed to pay the issue price of $1,000 for the bond. The curve in Figure 2-2

---

[39]   Recall that this was disputed by the utilities, but that the original PUC order was not protested on these grounds.

Figure 2-1
PAYOFF POSSIBILITIES ON A ONE-YEAR JUNK BOND

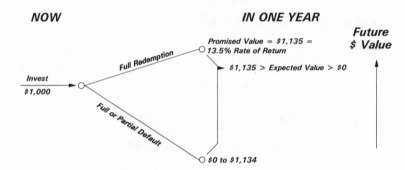

Expected Value Lies Between Promised Value and Zero,
and Depends on Odds of, and Amount Recovered in, Default.
Value of Bond Today = Amount Invested = $1,000.

depicts a hypothetical distribution of the firm's possible future asset values. Clearly, the bondholders will be paid the promised 13.5 percent only if the value of the firm at the end of the year is at least $1,135 per bond sold. If the value is less, the firm will default or renegotiate, and the bondholders will receive less than $1,135.

In this example, the expected value of the firm one year hence is $1,235 per bond, $100 more than bondholders have been promised. But the firm may be worth a lot less. If the realized value ends up to the right of the $1,135 point, the value of the firm is high enough to pay off bondholders and have money left over. But to the left of the $1,135 point, bondholders are not paid in full.

The area under the curved line between any two asset values represents the probability that the firm will have a future realized value in a particular range. So, the total area under the line to the left of the $1,135 point, the shaded region, is the probability that bondholders will receive less than the promised rate of return. In Figure 2-2, we assume this

Figure 2-2

POSSIBLE VALUES FOR THE FIRM AT JUNK BOND MATURITY DATE

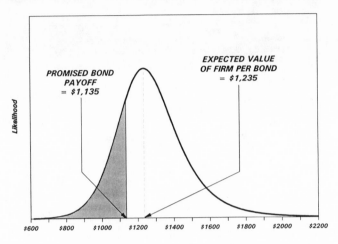

probability is 27 percent. This shaded area is the only range of relevance to bondholders: whether the firm is worth $1,136 or $100,136 per bond, bondholders get no more than $1,135. Thus the distribution of possible realized returns to bondholders is *asymmetric*, skewed to the left in the figure.

Figure 2-3 plots this area of possible realized returns to bondholders. It depicts an asymmetric distribution of payoffs to bondholders ranging up to $1,135. There is a 73 percent chance that bondholders will be given $1,135 and a 27 percent chance that they will receive something less. The 'something less' distribution depicted in Figure 2-3 has a mean of $1,055. Thus, the expected return to bondholders is 11.3 percent, not the promised 13.5 percent.[40] For comparison, the expected rate of re-

---

[40]   That is, 0.73 x $1,135 + 0.27 x $1,055 = $1,113. Therefore, the expected rate of return is: [(1,113/$1,000) - 1.0)] = 11.3 percent.

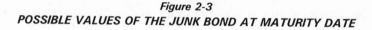

Figure 2-3
POSSIBLE VALUES OF THE JUNK BOND AT MATURITY DATE

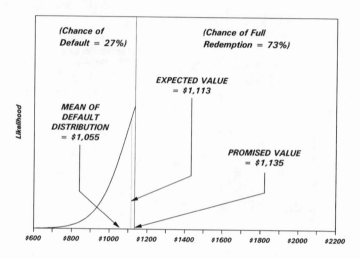

turn on a one-year, default-risk-free Treasury note might be 8 or 9 per-cent.[41]

Is 11.3 percent 'fair,' given the risk? It is if bondholders willingly paid $1,000 for the bond in the first place.[42] If the risk is too great at only an 11.3 percent expected rate of return, however, investors will offer less than the $1,000 issue price of the bond. This raises the promised

---

[41]   These yields are intended only to be illustrative, since the relationship among inter-est rates of different risk and the level of interest rates changes materially as eco-nomic conditions change. For example, promised junk bond yields were in the 15-16 percent range for much of 1990, while seven-year Treasury securities were in the 8-9 percent range. But after the Iraqi invasion of Kuwait, junk bond yields began to climb sharply, to 20 percent by mid October, even though seven-year Treasury yields remained below 9 percent (*Wall Street Journal* 1990, C1).

[42]   In the remainder of the chapter we use "fair return" in this sense. A fair return thus implies that the present value of the future cash flows that investors expect from the investment equals the amount they initially invested.

rate of return above 13.5 percent and the expected rate to a level investors consider fair given the risk. For example, if investors are only
willing to offer $980 for the claim on the payoff possibilities depicted in
Figure 2-3, the promised rate of return rises to 15.8 percent and the
expected rate of return to 13.6 percent.

The discussion above should help to distinguish the fair allowed rate of
return under traditional rules of rate regulation from the fair allowed rate
of return under the rules endorsed in *Duquesne*.

Regulatory Risk

The traditional paradigm of rate regulation (described in the Introduction
section of this chapter) came under pressure from forces such as oil price
shocks, double-digit inflation, uncertain electric demand, and changing
and retroactive safety standards for plants under construction, particularly
nuclear plants. The *average* cost of electricity went higher, and the
range of possible outcomes widened.[43]

As a result of these forces, regulatory rules often changed. Regulators
interpreted prudence standards to mean that economic disasters required
'sharing' of risks between investors and ratepayers (Tye 1989, 17).
They also disallowed unused and unuseful investments, however prudently incurred, from the rate base and in some cases forced investors to bear
the amortization costs as well.[44] As *Duquesne* also illustrates, in some
cases these decisions involved retroactive changes in regulatory rules that
shifted more risk to investors after the cards were face up and sometimes

---

[43]   For a survey of the effect of these developments on the industry, see James Cook
(1985, 82).

[44]   For an extensive survey of these cases, see Edward Berlin (1984, 26) and Richard
J. Pierce (1984, 497). For examples of the large disallowances *In re Public Serv.
Co.*, 130 N.H. 265, 539 A.2d 263 (1988), *appeal dismissed*, 109 S. Ct. 858 (1989);
Cleveland Elec. Illuminating Co. v. Public Util. Commission of Ohio, 4 Ohio St. 3d
107, 447 N.E.2d 746, *appeal dismissed*, 464 U.S. 802 (1983); *Office of Consumers'
Counsel v. Util. Commission*, 67 Ohio St. 2d 153, 423 N.E.2d 820 (1981), *appeal
dismissed sub nom.*, *Cleveland Elec. Illum. Co. v. Office of Consumers' Counsel*,
455 U.S. 914 (1982).

without specific regard to whether such risks were explicitly contemplated by the rate of return methodology applied prior to the switch.

The critical point is that the rate base, previously a largely known quantity not subject to significant risk, became a highly uncertain quantity, and the direction of movement in asset values was strictly downward. Under the new rules, investors might not be allowed to earn a return on all of the capital invested, because some of the investment might not be allowed in the rate base. Good outcomes, however, were still treated under the old rules: investors expect a return equal to the cost of capital. Thus, the new rules of rate regulation apparently approved by the *Duquesne* decision create asymmetric returns for utility investors analogous to those for junk bondholders.[45]

The result is asymmetric regulation, with the type of payoff structures depicted in Figure 2-4. If everything goes as planned, investors expect the full investment to be put in the rate base and to earn an exactly fair return, so the value of the plant to them as of today equals the amount invested.[46] (A familiar theorem of rate regulation says that if regulators set the allowed rate of return so investors expect to earn the cost of capital on the approved rate base, the value of the firm will equal the value of the rate base.[47]) But if some of the investment is disallowed, they expect a lower return. If in one outcome the plant is worth exactly

---

[45] As discussed below, regulatory institutions have long embodied a certain degree of asymmetry in returns. For example, the Court in *Duquesne* found that the modified prudent investment standard was previously in effect. The issue under traditional regulatory rules is whether these risks were *material*. In this part of the chapter, we assume the initial rules are symmetric to simplify the exposition.

[46] In Figure 2-4, and later in Figure 2-7, we make use of the most basic valuation formula: the value today of an expected cash flow to be received one year hence equals the expected cash flow divided by one plus the cost of capital. More elaborate present value formulas might be substituted for this simple one, but their use would lead to exactly the same conclusions. See generally Brealey and Myers (1988, Chapters 5 and 6).

[47] See Myers (1972, 73-76) for a discussion of this theorem and its implications. See Kolbe, Read and Hall (1984, 20-33) for a proof and a discussion of the component concepts.

what it cost and in the other it is worth less to investors than it cost, on average it will be worth less to investors than it cost.

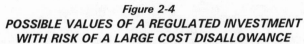

Figure 2-4
POSSIBLE VALUES OF A REGULATED INVESTMENT
WITH RISK OF A LARGE COST DISALLOWANCE

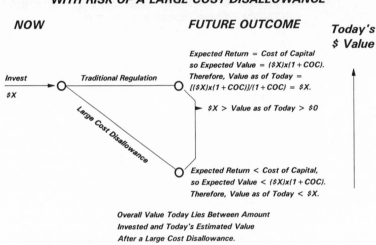

Figure 2-4 is the embodiment of the point we are making, and many of the objections we have received to earlier drafts of this chapter, which we discuss in Chapter 3, stem from a failure to appreciate the elementary logic of Figure 2-4. The point could not be more basic: if in one outcome investors get a payoff equivalent to just getting their initial money back, which by the simple mathematics of Figure 2-4 is true *regardless* of the magnitude of the cost of capital, and in the other outcome they get less, they face a losing game.

The actual outcomes in Figure 2-4 are stated in terms of *expectations*. Therefore, by definition there is a chance investors will get more or less than they expect on the upper branch, under the traditional regulation outcome. This merely reflects the fact that even under traditional regulation, there are always deviations of realized values from a precisely cost-based rate in practice. For example, rates are set periodically and inves-

tors are free to win or lose during the 'regulatory lag' between hearings.[48] Thus the value of a plant to investors upon completion may turn
out to differ somewhat from its cost.

Figure 2-5
POSSIBLE VALUES OF A POWER PLANT UPON COMPLETION
UNDER TRADITIONAL REGULATION

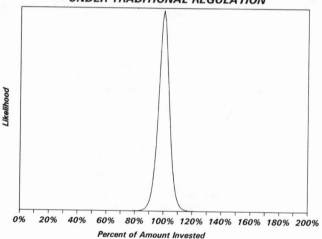

The existence of different possible realized asset values under traditional
prudent investment regulation, which changes none of our conclusions,
is shown in Figure 2-5. Figure 2-5 depicts investors' uncertainty today
about how valuable the investment may actually turn out to be in the
future, when it is completed. That is, a plant may turn out to be worth
more or less than its cost due to normal variations in the way regulation

---

[48]   Myers (1972, 83-84) advocates conscious use of regulatory lag as an incentive for
cost efficiency.

is applied.[49] Note that for the purpose of illustrating the relevant principles, Figure 2-5 initially makes the common assumption that there is a symmetric distribution of future realized asset values around the cost.[50] This is based on the assumption of a regulated monopoly where there is no bias in actual earnings either upward or downward from the allowed rate of return.[51] This simplifying assumption is relaxed below to consider the more general case of asymmetric distributions of realized values even under a prudent original cost standard.

If a rate base disallowance occurs, the range of outcomes is wider and the plant is worth less than it cost on average. Figure 2-6 adds an illustrative distribution of possible outcomes with disallowances to the tradi-

---

[49] We have made no assumption about just how this distribution of values is generated. A standard approach would be to assume the value will be given by the expected future investor cash flows upon completion, discounted by the then-current cost of capital. Another, somewhat different shape for Figure 2-5 would result if we assumed that under traditional regulation, the range of values of the project was wide, but that ratepayers had a 'call option' on good outcomes and that investors had a 'put option' to ratepayers for bad outcomes. If the 'exercise prices' of the call and put were equidistant from the mean, a narrow, symmetrical, truncated distribution would result. For present purposes, however, we need not speculate on the 'true' shape of Figure 2-5 or the 'true' determinants of the cash flow distributions or of the cost of capital. We need only note that investors today cannot know for sure whether the plant under construction will be worth more or less than it cost when it is complete, and that for utilities, the range of uncertainty under traditional regulation was believed to be rather narrow, compared with that for unregulated firms.

[50] The term 'symmetry' has two possible interpretations: (1) the *distribution* is symmetric, that is, the shape on one side of the mean is the mirror image of that on the other side; and (2) the *payoff to investors* is symmetric, that is, the probability-weighted average of the possible values above the cost of the plant equals the probability-weighted average of the values below the cost of the plant. As drawn, Figure 2-5 satisfies *both* interpretations. When we intend the first sense of the term, we refer to symmetric or asymmetric *distributions* of possible payoffs. When we intend the second sense, we refer to symmetric or asymmetric *payoffs*.

[51] It might at first seem more natural to think in terms of variation in earned rates of return from the cost of capital, rather than of plant values as a percentage of cost. But plant value embodies the impact of expected deviations between the earned rate of return and the cost of capital over the life of the plant. Thus, it is a more complete measure.

tional case of Figure 2-5. There is a material chance the plant will be worth nothing and little chance it will turn out to be worth what it cost.[52] What this means for investors depends on how great the risk of a disallowance is. For example, Figure 2-6 also shows the plant value distribution assuming a 20 percent chance of a disallowance,[53] with an expected value of 87 percent of cost, not 100 percent.[54] Under any such set of asymmetric expectations, investors would expect future cash flows worth less than the amount of money they invested.[55]

---

[52]   The distribution is drawn so the expected value of the plant, if there is a disallowance, is about 35 percent of its cost.

[53]   Figure 2-6 assumes that the probability distribution of disallowance possibilities is known. To employ a distinction first made by Frank Knight (1921), this is a case of *risk*. The actual situation is more appropriately described as *uncertainty*, in Knight's terms, because investors do not know the probability distribution of disallowance possibilities.

[54]   The shape of the 'disallowance' distribution is purely illustrative. The value to shareholders might even be negative, if the amounts owed the plant's bondholders are greater than the value of any tax write-offs associated with the disallowance. Further, we assume the value distributions are independent of one another, that is, that knowledge of the value that this particular plant would have under traditional regulation does not provide information about the value it would have if there were a disallowance. (This is unrealistic in practice; e.g., low-cost plants would probably be exposed to smaller disallowances.) Given the independence assumption, the mean of the value distribution that assumes a 20 percent chance of disallowance equals the weighted average of the means of the two starting distributions: (0.8 x 100%) + (0.2 x 35%) = 87%.

[55]   The D.C. Circuit recognized a similar phenomenon in a proposed Federal Communications Commission rule that would have refunded telephone company earnings above the allowed rate of return. It stated

> The refund rule requires the carrier to refund any earnings above the upper bound of [the allowed rate of return plus a small amount], while the carrier may not recoup any shortfall in its earnings below [the allowed rate of return]. . . . Thus, over the long run the carrier is virtually guaranteed to fall short of earning its required target rate of return. . . . [S]ince the Commission views the rate of return as a minimum [necessary return], the refund rule under the Commission's view would operate over the long run to put a carrier out of business. It should be stressed that this result does not reflect merely the business risk that a carrier is bound

Figure 2-6

**DISTRIBUTION OF POWER PLANT VALUES UPON COMPLETION FOR A
0%, 20% AND 100% CHANCE OF A RATE BASE DISALLOWANCE**

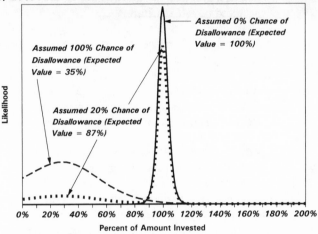

Here we define *regulatory risk* as the risk due to an asymmetric distribution of possible plant value outcomes.[56] Thus, the key economic ques-

---

> to accept under the accepted view that regulation does not guarantee the regulated company a profit. Rather, it is the Commission's refund rule that seems to guarantee the regulated company an economic loss.
>
> *AT&T* v. *FCC*, 836 F.2d 1386, 1390-1391 (D.C. Cir. 1988).

The situation in the telephone case can be interpreted as a truncation of the right side of the 'traditional regulation' value distribution in Figure 2-5 -- similar to that which occurs for bondholders in Figure 2-2 -- which results in an expected rate of return below the promised rate of return. The D.C. Circuit overturned the proposed rule due to inconsistency with the FCC's own theory of rate of return regulation.

[56] There appears to be no generally accepted definition of regulatory risk. Chang Mo Ahn and Howard E. Thompson (1989), for example, appear to define regulatory risk as the impact of regulation on the cost of capital. That is explicitly *not* our definition.

tion posed by *Duquesne* is how investors can be appropriately compensated for such risks. Four possible answers are:

1.   Increase the allowed return on investment to an amount greater than the cost of capital, by addition of a "regulatory risk premium" to restore a balanced payoff structure;[57]

2.   Eliminate asymmetric payoff distributions by changing regulatory practices;

3.   Add a compensating cost-of-service item, akin to a fee or insurance premium for provision of a risky service, to the revenue requirement; and

4.   Adjust another cost recovery item by an amount sufficient to offset the asymmetry.

For the bulk of this chapter we will focus on the first method, the solution the Supreme Court presumed operative in the case of *Duquesne*. It is doubtful that this presumption was specifically true for *Duquesne* and questionable that it can work in situations of considerable regulatory and other asymmetric risk. In the final section and in Chapter 3, we expand on the possibilities for other solutions.

Figure 2-7 depicts one method of compensation for regulatory risk. In this case, investors are allowed a rate of return above the cost of capital if the plant does not experience any cost disallowance as compensation for the risk that it might. If the regulatory risk premium is set appropriately, investors can expect, on average, that the value of the plant will equal their investment. However, individual plants may turn out to be

---

[57]   As noted above, promised investor returns equal the allowed rate of return times the rate base. Compensation for regulatory risk could be accomplished through an addition to the rate base. However, such a mechanism would almost surely involve a rate of return premium that subsequently would be capitalized in the rate base (e.g., a higher rate of return for the 'allowance for funds used during construction'). Thus, to focus on risk compensation through a rate-of-return premium rather than a rate base adjustment is not a restrictive assumption.

worth materially more or less. We identify this case as providing a 'compensated' risk of a disallowance.[58]

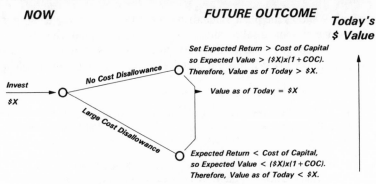

Figure 2-7
POSSIBLE VALUES OF A REGULATED INVESTMENT
WITH COMPENSATED RISK OF A LARGE COST DISALLOWANCE

The distribution of possible plant values for a compensated 20 percent chance of a disallowance is illustrated in Figure 2-8. The 'no-disallowance' distribution of Figure 2-6, but not the disallowance distribution,

---

[58] We refer to a 'compensated' chance of a loss in the same sense as other articles address compensated transitions among rate base methodologies (Myers, Kolbe and Tye 1985, 111-112; Myers, Kolbe, and Tye 1984, 42-43) or among regulatory treatments of deferred income taxes (Kolbe, Tye and Baker 1984, 442-445).

has been shifted to the right by the regulatory risk premium.[59]   The distribution that assumes a 20 percent chance of disallowance also shifts right, such that its expected value becomes 100 percent.[60]   Note that the odds of achieving this value on any given plant are virtually zero: almost all plants either 'win' or 'lose' by substantial amounts, once they are built, and the outcome is fair (equal to 100 percent) only on average.

Figure 2-9 shows the plant value distribution for a compensated 50 percent chance of some disallowance. In this case, the rate of return has to be set high enough so the plant's worth to investors is more than one and one-half times its cost if it is put into the rate base without any disallow-

---

[59]   The regulatory risk premium proposed herein has legal precedent in decisions preceding *Duquesne*. See *Washington Gas Light* v. *Baker*, 188 F.2d 11, 19 (D.C. Cir., 1950), *cert. denied*, 340 U.S. 952, 71 S.Ct. 571:

> Here, the Commission adopted the prudent investment theory of rate base valuation rather than the reproduction cost method. Appraisal of the former theory reveals that the "used and useful" standard is no necessary part of it. Primary emphasis is now being placed not on "specific property, tangible and intangible," but on capital prudently invested and embarked on an enterprise in the public service. Under the view taken by the Commission, the theory contemplates that rates will enable the investor to maintain his original prudent investment intact until it is recovered through annual charges to depreciation expense, which are reflected in a reserve for depreciation.
>
> *   *   *
>
> [T]he rate of return to which the investor is entitled is measured in part by the risks of the business as compared with those of comparable enterprises. It thus becomes relevant to determine whether or not investors have, during the useful life of this property, been compensated for assuming the risk that it would become inadequate or obsolete before the investment in it was entirely recovered. Such compensation may have been made either through inclusion of obsolescence (1) as one of the elements used in calculating depreciation expense, or (2) as a risk considered in fixing the permissible rate of return. If, in the past, the risk of obsolescence was provided for in either of these two ways, then the abandoned property should not be included in the rate base today. 188 F.2d 19, 20 [footnotes omitted.]

[60]   Thus, Figure 2-6 represents both an asymmetric distribution and an asymmetric payoff to investors, while Figure 2-8 -- despite its asymmetric distribution -- represents a symmetric payoff in the sense we defined above.

Figure 2-8
DISTRIBUTION OF POWER PLANT VALUES UPON COMPLETION FOR A
COMPENSATED 20% CHANCE OF A RATE BASE DISALLOWANCE

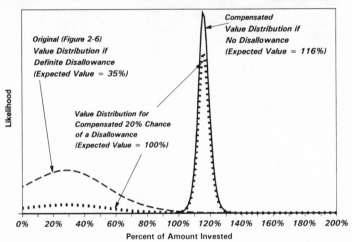

ance. This creates serious practical difficulties addressed later in the chapter.

Two Classes of Regulatory Risk

To understand the economics of *Duquesne*, a final distinction must be made between two classes of regulatory risk. The first is that just discussed: the risk of some disallowance of the invested capital from the rate base or other event that negatively skews the distribution of returns within a consistently applied ratemaking methodology. Note that as the Court recognized,[61] an adjustment to the allowed rate of return (or equivalent compensation) is required even in this case.

---

[61] *Duquesne*, 109 S.Ct. at 617-18.

Figure 2-9

DISTRIBUTION OF POWER PLANT VALUES UPON COMPLETION FOR A COMPENSATED 50% CHANCE OF A RATE BASE DISALLOWANCE

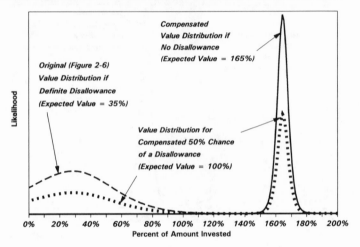

The second is a retroactive shift in the distribution of possible disallow-ances due to a change in regulatory oversight, the issue that brought this case before the Court. This is depicted in Figure 2-10, which postulates that the compensated 20 percent loss case of Figure 2-8 has been in effect,[62] but that a new, more severe range of possible losses under the disallowance case is established by a regulatory rule change. Even if the no-disallowance distribution remains at the point where it provided fair compensation under the old rules, it is now inadequate under the new rules: the new value distribution with a 20 percent chance of a disallow-ance has an expected value below 100 percent (96 percent in this exam-ple). Investors again lose on average.

---

62   This is an economic interpretation of the Court's presumption in *Duquesne*.

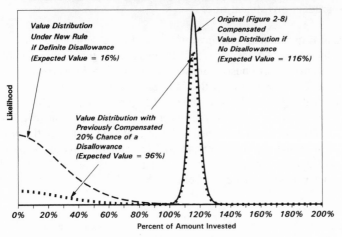

Figure 2-10

**DISTRIBUTION OF POWER PLANT VALUES UPON COMPLETION
FOLLOWING RULE CHANGE WITH PREVIOUSLY COMPENSATED
20% CHANCE OF SOME RATE BASE DISALLOWANCE**

Regulatory Risk Under the Prudent Investment Rule

Figure 2-5 assumed that the distribution of possible future realized outcomes prior to the introduction of the risk of disallowance of prudently
incurred, but unused and unuseful, investment was symmetric. Comparing Figure 2-5 with Figure 2-6, we see that a symmetric distribution
exists under traditional regulation only if the risk of a disallowance under
the pure prudent investment standard is zero. That is, an asymmetry like
that in Figure 2-6 could result as well if the investment were found to be
imprudent, a possibility even under the original prudent investment
test.[63]

The possibility of a nonzero probability of a finding of imprudence thus
questions the fundamental assumption that underlies economists' advice

---

[63] Several readers of earlier drafts, including Alan Buchmann and Ernest Ellingson,
have suggested the traditional rules in at least some jurisdictions included the risk
of asymmetric losses.

to equate the allowed rate of return with the cost of capital.   If this
assumption does not correspond to legal reality, economists have been
giving bad advice.[64] As long as there is a nonzero probability of disallowance, the economic prescription to equate the allowed rate of return
with the cost of capital apparently needs to be changed even under the
pure prudent investment rule.

A natural question is why investors would have supplied capital if this
were the case. An answer we find persuasive is that the probability of
disallowance of an unused, unuseful or imprudent investment was perceived to be *de minimis*. If this is in fact the case, the interpretation of
'traditional regulation' in the context of Figure 2-4 and Figure 2-5 is that
the probability of -- and/or the expected loss in the event of a finding
of -- imprudence was perceived to be so low as to be lost in the noise of
the error in the estimation of the cost of capital. This fortunate result
can no longer be assumed to prevail in today's regulatory climate, where
asymmetry now must be confronted head on.[65] For present purposes,

---

[64]   Myers (1972) speaks only of setting rates so investors expect to earn the cost of
       capital, not of equating the allowed rate of return to the cost of capital. However,
       by the time of, for example, Kolbe, Read and Hall (1984), it was common to as-
       sume Myers's prescription would be achieved if the allowed rate of return were
       equated to the cost of capital, providing that the other costs of service were estimat-
       ed accurately. This chapter shows that this common assumption is unrealistic.

[65]   It is difficult to disagree with the conclusion that the degree of regulatory risk has
       changed substantially in the 1980s:

       When I researched this topic for other purposes in 1983, I conducted an
       exhaustive search for regulatory disallowances based on imprudence. The
       Federal Energy Regulatory Commission (FERC) and its predecessor,
       FPC, had never disallowed an investment on the basis of imprudence in
       the agency's fifty-year history. [footnote omitted] I could find only a few
       cases in which state agencies had disallowed investments based on a find-
       ing of managerial imprudence. Even in those rare cases -- about one per
       decade -- the magnitude of the disallowance was relatively trivial. [foot-
       note omitted] The aggregate amount disallowed in the history of utility
       regulation probably did not exceed a few hundred million dollars. By
       contrast, during the period 1984 through 1988, state agencies disallowed
       as imprudent significant portions of the investments in nineteen completed
       generating plants. The average amount disallowed per plant was $610
       million; the aggregate amount disallowed was $11.6 billion. [footnote

for simplicity, if we redefine what we mean by 'traditional regulation' to include a negligible risk of material asymmetric losses, our analysis is unchanged.

In actual situations there may be asymmetry arising from economic forces as well as regulatory actions. For example, where the firm may be subject to both regulatory and competitive constraints, the distribution may be skewed downward, and the expected return falls below the allowed return. This was the case in *Market Street Railway Co. v. Railroad Commission of California*,[66] where the Court ruled that it was market forces rather than regulatory restraint that prevented the firm from earning its cost of capital.[67] Another example might be an environmental cleanup liability that regulators refused to pass through to customers.[68] These situations, in practice, would have to be corrected by a mechanism similar to that proposed here for regulatory risk of disallowance of unused and unuseful investments.

---

omitted] If these agency findings are to be believed -- that is, if the findings of the past four years are something other than a guise for politically opportunistic exercises of raw political power to redistribute wealth from a minority to the majority -- then they suggest a startling trend in the industry's management. Apparently, for decades electric utility managers were almost uniformly individuals with outstanding business acumen. At some point in the 1980s, this entire generation of exceptional managers was replaced en masse by a generation of bumbling idiots. (Pierce 1989, 2031, 2050-2051)

One is reminded of Mark Twain's surprise that his father, having been an ignoramus at Twain's age of 14, could have learned so much in the 7 years it took Twain to become 21.

[66]   324 U.S. 548 (1945).

[67]   *Id.* at 566-67.

[68]   See William W. Hogan and A. Lawrence Kolbe (1991) for a discussion of proposals to disallow potentially hundreds of millions of dollars of environmental cleanup costs in a generic proceeding before the Massachusetts Department of Public Utilities. The risk exposure represented a large share of the net worth of the local gas distribution companies in the Commonwealth.

What Type of Rate of Return on Equity
Is Typically Allowed by Regulators?

To this point we have implicitly assumed that regulators ordinarily equate
the allowed rate of return under the old prudent investment rule with the
expected rate of return on equity, not with the equivalent of a *promised*
rate of return on equity as defined in the bond example. The basis of
our assumption is the definition of the cost of capital given above: the
expected rate of return in capital markets on alternative investments of
equivalent risk (Brealey and Myers 1988, Chapter 7; Kolbe, Read and
Hall 1984, Chapter 2). The term 'expected' in this definition is precisely
the same as that defined in the junk bond analogy. Indeed, the cost of
capital of a junk bond *is* its expected rate of return, not the promised
yield to maturity that is ordinarily quoted (Brealey and Myers 1988, 561-
562).

Nonetheless, when regulators set the allowed rate of return on debt, they
use the promised interest rate, which produces the promised yield at the
time the debt is issued, not the expected rate.[69] No reduction in the
promised amount is made for the fact that bondholders expect somewhat
less because the firm may someday default on its debt. There are at least
two good reasons for this. First, the promised rate on debt is all that can
be directly observed; calculation of the rate bondholders expect would be
difficult at best. Second, bondholders are treated fairly if they are paid
the full promised rate (13.5 percent in the above example). If they were
paid only the expected rate (11.3 percent) without default so that 11.3
percent became the upper limit, the true rate they would expect, given
the risk of default, would fall well below 11.3 percent; bondholders
would be shortchanged.

The expected and promised rates on bonds differ only because of the
asymmetric distribution (in the first sense of the term defined above) of
bond payoffs. Regulatory rules that create asymmetric distributions for
investors can make the expected and 'promised' rates of return differ for
equity as well. However, unlike debt, *the 'observed' rate of return for*

---

[69] Regulation typically allows a rate of return on the debt share of the rate base equal
to the embedded interest rate on outstanding debt.

equity is its expected rate of return, not the equivalent of a promised rate of return.[70] Yet in the presence of an asymmetric distribution, regulators have to allow the equivalent of a promised rate of return on equity if investors are to have a fair chance to earn the expected rate of return as measured by testifying experts.[71]

ECONOMIC IMPLICATIONS OF *DUQUESNE*

The previous section leads immediately to an economic interpretation of the Court's view of the situation in *Duquesne*.

As a rule, the computation of the allowed rate of return must be consistent with the regulatory risks inherent in the regulatory system used. In the Court's view, Duquesne Light's investors were on a regulatory system something like that of Figure 2-8 or Figure 2-9, with an allowed rate of return that compensated them for the risk of a lost return *on* (but not

---

[70] There are a variety of methods to estimate the cost of equity capital, but all aim at the expected rate of return (Kolbe, Read and Hall 1984, Chapter 3). The intuitive way to see this may be to suppose the rate of return analyst in a particular case had a perfect sample, a group of utilities that had already had past disallowances and still were at risk of having future disallowances with the exact frequency distribution facing the utility in the above examples. Then any commonly used method of estimating the cost of equity capital, whether based on a backward look at history or a forward looking interpretation of current stock market evidence, would measure the *expected* rate of return for the sample, which would be the expected rate of return for the company in question as well. For example, in Figures 2-8 or 2-9, a drawing of a large sample of firms facing either figure's distribution of returns would produce a rate of return consistent with the expected value, not the 'promised' (i.e., *allowed*) rate of return on plants that suffered no disallowance.

[71] To prevent misunderstanding, we will note that we are *not* recommending a guaranteed rate of return on equity in the absence of a disallowance. The equivalent of a 'promised' rate of return on equity is merely a rate of return equal to the cost of equity capital plus a premium for the risk of a disallowance, just as the promised rate of return on a corporate bond equals the cost of capital for the bond plus a default premium. Utilities would no more be guaranteed to earn that rate of return in the absence of a disallowance than they are when the allowed rate of return equals the cost of capital in the absence of a disallowance.

of) capital should the CAPCO plants turn out not to be used and use-ful.[72] The Court suggests that regulatory systems that do not compen-sate for the regulatory risks inherent in that system might violate the Constitution,[73] but the used and useful standard, at least, does not do so intrinsically in and of itself.[74]

The Pennsylvania legislature after the adverse outcome was known im-posed a full used and useful test that denied investors the return of capital as well. The Court held that this is not inherently a taking of proper-ty,[75] if offset by other factors,[76] but can raise "serious constitutional questions" otherwise.[77]

The opinion cites no evidence of other offsetting factors, so the result is a situation like that in Figure 2-10 that imposed a windfall loss on Duquesne Light's investors. If the switch had been severe enough, the new expected rate of return after the switch would have fallen outside the normal range of variation of regulatory outcomes[78] and, according to the Court's analysis, issues of constitutional taking would exist.[79]

But despite the apparent lack of other offsetting factors, the loss as mea-sured by the Court was so slight that it fell within the normal range of

---

[72]   *Duquesne* at 617-18.

[73]   *Id.* at 619.

[74]   *Id.* at 612.

[75]   *Id.*

[76]   *Id.* at 617, 619.

[77]   *Id.* at 619.

[78]   We presume this means outside the narrow, symmetric band of possible realized outcomes depicted within the traditional regulation curves in Figure 2-5, but we raise this question again later.

[79]   *Duquesne* at 619.

variation of approved regulatory outcomes. Therefore, constitutional taking issues did not arise.[80]

The implication is that the Court found no constitutional distinction between a bad outcome realized under a system where good outcomes balance bad ones on average, and a rule change that always imposes a certain loss, without any offsetting chance for a gain, as long as the loss is 'slight.' If an adverse, retroactive switch in ratemaking methodology produces a capital loss that conceivably could have occurred in the prior regulatory regime, the end result will be found constitutional under this test. Rational investors, however, will recognize the difference between a random event that could happen in a fair game and the results of a biased game in which adverse changes are intentionally imposed. This recognition leads to the problem of how to compensate investors for this risk.

In the next section, we show that these findings create a regulatory system that may prove difficult or impossible to implement fairly. First, we discuss measurement problems for attempts to arrive at a fair promised rate of return on equity. Then we show that the very need to use a promised rate of return makes that rate of return even harder to determine fairly. Finally, we show that the Court's definition of slight is biased downward, which makes these problems still more severe in practice.

Measurement Problems for the Allowed Rate of Return under Asymmetric Regulation

Regardless of what has been true historically, *Duquesne* makes clear that certain rate base methodologies, consistently applied, have regulatory risks for which investors must be compensated in the allowed rate of return. The Court's own ranking of the risks inherent in the alternative rate base/rate of return standards demonstrates that recognition of the regulatory risk inherent in the modified prudent investment and used and useful standards is required.

---

[80]   *Id.* at 618.

The opinion therefore requires the development of improved procedures for incorporating regulatory risk into rate base/rate of return methodologies. We need to find ways to infer the equivalent of a promised rate of return on equity in order to give investors a fair opportunity to earn the expected rate of return on equity calculated with the standard methods.[81]

Unfortunately, it is difficult to imagine how this can be done rigorously. Unlike debt, equity contains no well-defined claim on a set of promised cash flows that, with the addition of the original amount paid for the security, would permit computation of a promised yield to maturity on equity. The figures above simply assume we know the distribution of values investors expect under a disallowance case to illustrate the principles. The data necessary to construct such figures in an actual case are almost certain to be difficult to identify.

Despite this fact, *Duquesne* teaches that regulated firms that make investments in jurisdictions with rules that produce asymmetric distributions of outcomes now need a regulatory risk premium over and above the estimated cost of equity capital.

Calculation of a Regulatory Risk Premium

Given the absence of data, some commissions may simply choose to add a risk premium without detailed analysis. There is no way to tell if such a premium is truly compensatory, however. A more careful approach would be to make an explicit assessment of the expected loss if there is an event with an asymmetric effect and of the likelihood that such an event will occur. Given estimates of such losses, how should a commission that wants to make an explicit assessment proceed?

The answer depends in part on how the regulatory risk premium is to be implemented. As an example, we consider the possibility that a plant under construction will be later declared unused and unuseful. The most

---

[81]   See Kolbe, Read and Hall (1984, Chapters 3 and 4) for discussion of these standard methods.

direct approach would be to add a premium to the allowance for funds used during construction (AFUDC) rate.[82] The premium will be capitalized with the rest of AFUDC if the plant goes into the rate base, thereby providing a fair promised return for the risk during construction.

The fair promised AFUDC rate for a given year could be derived with the following logic. Construction amounts expended by the start of the year, including past accrued AFUDC, need to have an expected value equal to the initial amount plus the return on capital at the end of the year.[83] The expected value is a weighted average of the promised return times the probability that the plant is still on track at the end of the year plus the disallowance return times the probability the plant has been abandoned. Mathematically, this is as follows:[84]

$$K_0 \times (1 + coc) = K_0 \times (1 + aror) \times (1 - Pr_d) + K_0 \times (1 + disr) \times Pr_d \quad (1)$$

where

K_0 = the cumulative capital outlays as of the start of the year, including previous AFUDC;

coc = the cost of capital;

$aror$ = the allowed (i.e., the promised) rate of return used for AFUDC;

---

[82] Under AFUDC, interest and return on equity for construction in progress is accrued and capitalized during the construction period, but no cash payments are made by ratepayers until the construction is completed. Unless recovery is denied, the investment plus accrued AFUDC goes into the rate base at that time. An alternative to including the risk premium to the rate of return on AFUDC would be to allow a premium in the rate of return on total assets currently in service. This would avoid some of the difficulties discussed below with an AFUDC premium, but would seem contrary to the spirit of a used and useful statute. This alternative also would not be feasible for a stand-alone project.

[83] Amounts spent during the year can be treated exactly the same way, but with rates of return prorated for the partial year.

[84] With appropriately defined variables, Equation (1) and those that follow could be used either with the overall rate base and allowed rate of return or with the equity rate base and the allowed rate of return on equity.

Pr_d = the probability that an event that triggers the disallowance return will occur over the next year; and

disr = the disallowance return, which is usually negative.[85]

Equation (1) can be solved for the fair allowed rate of return for AFUDC:[86]

$$aror = \{[(1 + coc) - (1 + disr) \times Pr_d]/(1 - Pr_d)\} - 1 \qquad (2)$$

For example, suppose a commission believes that there is a 10 percent chance of plant abandonment during the next year and that the company's cost of equity capital is 15 percent. Suppose the company is on a pure used and useful system, and that if the plant is abandoned shareholders lose the entire investment.[87] Under this procedure, disr = -1 because

---

[85] We are indebted to Nancy Lo for pointing out that Equation (1) as specified combines the allowed rate of return in the event of a disallowance and the percentage magnitude of the disallowance into a single term, disr. A somewhat different formula would result if these two components of disr were shown as separate terms. In actual applications, there may be times when it is more convenient to separate these two terms.

[86] If it were more convenient to separate term disr in Equation (1) into disr = $disr^*$, the (negative) magnitude of the disallowance, + aror, the allowed rate of return, and if the allowed rate of return on the approved rate base were the same regardless of whether a disallowance occurred, Equation (2) would be given instead by

$$aror = \{(1+coc)/[1+(Pr_d)x(disr^*)]\} - 1 \qquad (2^*)$$

In any actual application, it would be important to keep track of the distinction between $disr^*$, the magnitude of the disallowance itself, and disr, the overall rate of return on the initial investment, K_0, in the event of a disallowance.

[87] The actual shareholder loss is uncertain. Plant abandonments may generate tax benefits even under a pure used and useful rule. These benefits might in principle be passed through to ratepayers, leaving shareholders with no tax benefit. It seems more likely, however, that the saved income due to the tax deductions would be left to investors on the ground that the tax losses correspond to assets that are not used and useful. The present value of future tax benefits then would partially offset the

investors lose 100 percent of their investment to date, and the formula for the fair AFUDC rate is

$$aror = \{[(1 + 0.15) - (1 - 1) \times 0.1]/(1 - 0.1)\} - 1$$
$$= (1.15/0.9) - 1$$
$$= 27.8\%$$

Suppose instead the company had been on the modified prudent investment standard and got back payments with a present value of 50 cents on the dollar in the event abandonment occurred, so disr $= -0.5$. Then the fair allowed AFUDC rate would be

$$aror = \{[(1 + 0.15) - (1 - 0.5) \times 0.1]/(1 - 0.1)\} - 1$$
$$= [(1.15 - 0.05)/0.9] - 1$$
$$= 22.2\%$$

Thus, the Court's finding that the used and useful standard represents greater risk than the modified prudent investment standard is borne out by the simple economics of the fair rate of return. What may be surprising is the size of the regulatory risk premium required under the modified prudent investment standard. This regulatory risk premium is 7.2 percent (aror - coc $= 22.2\%$-15.0%), with only a 10 percent abandonment risk and 50 percent loss of capital given abandonment.[88] Table 2-1 shows the fair allowed AFUDC rates under a variety of disallowance probabilities and returns.[89] In practice, the high probabilities of aban-

---

loss. However, bondholders would still have to be repaid by shareholders out of other income, if the plant was not a stand-alone venture. Thus, shareholders might end up losing more or less than the amount of their original investment.

[88] Again, this -50 percent is the value of disr, not disr*. If the assumptions of the example were changed so that the -50 percent were the loss in K_0 and the allowed rate of return were granted despite this loss, a less severe penalty than assumed in the text, Equation (2*) would be relevant. In this case, aror $= 21.1$ percent, which still represents a 6.1 percent regulatory risk premium.

[89] If the potential loss does not affect the size of the rate base -- for example, if the loss is an unrecoverable environmental cleanup expense -- a formula different from either Equation (2) or (2*) holds. If L is the size of the potential loss, K is the rate base, coc the cost of capital, aror the allowed rate of return, and Pr the probability of a loss, the analogue of Equation (1) that defines the fairness requirement is [K

donment may be quite realistic. Even if the risk of abandonment were quite low in early years for plants that have actually been abandoned, for example, in later years they must have gotten quite high (approaching 100 percent at the end).

TABLE 2-1 FAIR ALLOWED RATE OF RETURN WITH CHANCE OF FULL OR PARTIAL DISALLOWANCE OVER THE NEXT YEAR				
Probability of a Disallowance (%)	*Rate of Return Given a Disallowance (%)*			
	-25%	*-50%*	*-75%*	*-10*
0	15.0	15.0	15.0	15.0
1	15.4	15.7	15.9	16.2
5	17.1	18.4	19.7	21.1
10	19.4	22.2	25.0	27.8
25	28.3	36.7	45.0	53.3
50	55.0	80.0	105.0	130.0
75	135.0	210.0	285.0	360.0

Notes: Assumes that the probability of a disallowance is independent of the fair allowed rate of return. Assumes the cost of capital is 15 percent.

---

x $(1 + coc)] = [K \times (1 + aror)] - (Pr \times L)$. This implies that the analogue of Equation (2) is aror $= coc + [(L/K) \times Pr]$. The fact that the rate base is not at risk reduces the size of the fair allowed rate of return, but the magnitude can still be far above rates of return normally allowed in rate cases. If there is a 50-50 chance of a loss equal to 50 percent of the rate base, for example, the fair allowed rate of return exceeds the cost of capital by 25 percentage points (that is, a 15 percent cost of capital becomes a 40 percent fair allowed rate of return).

We do not know if the PUC allowed a regulatory risk premium of the above magnitudes in the AFUDC rate for the CAPCO plants, but our experience suggests this is extremely unlikely.[90] In that case, the Court's presumption that the allowed rate of return included compensation for the probability of abandonment must rest instead on the presumption that the overall allowed rate of return on all of Duquesne Light's assets was set sufficiently above the cost of capital. The required premium in the rate of return on all assets would be smaller than that on the CAPCO assets alone, since it would be a weighted average of a (relatively large) premium on the CAPCO assets and no premium on the rest of the assets.[91] For example, since the CAPCO assets represented about 2 percent of Duquesne Light's rate base, a 50 percent regulatory risk premium on CAPCO[92] would require a 1 percentage point premium in Duquesne's *overall* rate of return.[93] The Court's presumption enabled it to avoid an investigation of whether such a premium in fact existed.

---

[90] A factor raising doubt that such was the case is the fact that Pennsylvania switched back to the modified prudent investment standard for all investments canceled after 1985. A cynical interpretation of this provision is that it was an effort to impose a retroactive switch on old investments without having to pay an additional regulatory risk premium on future investments under the used and useful test. Investors, unlike Charlie Brown in his belief that Lucy will not snatch away the football yet another time in the comic strip, may well be skeptical of attempts to switch methodologies only for sunk investments.

[91] This assumes the CAPCO plants in question were the only plants at risk of abandonment. If other plants under construction were also at risk, the required premium would be higher than calculated here.

[92] Table 2-1 suggests a 50 percent premium is not of an unreasonable magnitude in the last years of a troubled plant.

[93] Since Duquesne Light's common equityholders bear the loss, the risk premium should go to them. Duquesne Light's common equity ratio near the end of the CAPCO plants' abandonment was about 35 percent, so the regulatory risk premium on equity under these assumptions would be 1 percent/0.35, or almost 3 percentage points.

Interaction Between the Regulatory Risk
Premium and the Probability of Disallowance

If the probability of noncompletion is an objective number, the necessary adjustment is very simple. However, in practice, things might not be so simple. First, there is likely to be no objective, actual experience that would permit the determination of a consensus forecast for the probability of disallowance in a particular application. The assumed disallowance distribution in Figure 2-6 will vary greatly across investments and regulatory jurisdictions and is likely to be the object of great dispute in every rate case.

Furthermore, the examples in Table 2-1 show that the necessary allowed AFUDC rate could approach politically unacceptable levels in even relatively low-risk situations. And adding a premium for plants under construction to the overall allowed rate of return would carry its own set of legal or political problems. Is it a violation of a used and useful statute, for example?

Moreover, the risk of abandonment may not be independent of the allowed rate of return. During long lead times, the risk of abandonment may increase, necessitating increases in the cost of attracting more capital. This increases the cost of the project, leading to an even greater probability of abandonment. The system simply may not converge to a stable equilibrium with a politically acceptable rate of return and an economically viable probability of completion.

A policy of disallowance of unused and unuseful investments may also create perverse incentives for abandonment and continuation for ratepayers and investors. Investors may want to complete plants that it is socially rational to abandon, and ratepayers may want to abandon plants that it is socially rational to complete.

Given all of these circumstances, the Court's presumption that the allowed rate of return can be set to compensate for the regulatory risk may simply break down, particularly where the plant at risk is a large share of the total rate base: there may be *no* feasible fair allowed rate of return.[94] The only remedy if rate base/rate of return regulation is to be preserved in this case is to restore symmetry to the regulatory rules.

Moral Hazard

The odds of adverse rulings from events such as plant abandonment are not given solely by exogenous, economic factors, but rather in part depend on decisions of regulators themselves. This creates additional problems, akin to a set sometimes encountered in the insurance industry. When the party who benefits if an insurable risk occurs (the policyholder) is able to change the amount to be paid or the likelihood that payment will occur, 'moral hazard' is said to exist. Finding a fair insurance premium in the presence of moral hazard is difficult or impossible.[95]

The problem of moral hazard may be illustrated by a somewhat whimsical example. Suppose a world famous gunfighter invites a tenderfoot to a poker game, but reserves the right to pull out his gun and change the rules at any time. What up-front risk premium does the tenderfoot require if he is to join the game?

If the risk premium itself is not subject to expropriation via a later change in the rules, the required premium is the tenderfoot's maximum

---

[94] The situation is analogous to the practice in which a new junk bond contains a promise to adjust the interest rate so it trades at face value at a fixed future date. If the company is in enough trouble, raising the promised interest rate reduces the odds the promise will be kept, which requires a further increase in the interest rate, and so on. Western Union reportedly had a bond issue that it was unable to make trade at face value for this very reason (*Wall Street Journal* 1989a, C2; and 1989b, C17).

[95] Here we assume that there is no difference in the information available to the utility and the regulator, or at least that any differential information is swamped in importance by the true uncertainty about how much the plant will cost on completion and whether electric demand will have grown by enough to require the new capacity. We consider only the problem created by the fact that the regulator, having paid for insurance, is the party that makes the decision as to whether the insured event has occurred.

loss (i.e., all that he brings to the table). This is the only answer because the tenderfoot realizes the gunfighter can claim everything he brings to the table at any time.

Further, the gunfighter must at some point expropriate an amount equal to the risk premium to make the game fair: the gunfighter has already paid that amount to the tenderfoot as compensation for the right to change the rules. In this case the outcome is determinate: the risk premium is equal to the assets that the tenderfoot brings to the game, which are at some point seized by the gunfighter. (They may play a while first for the fun of it.)

If the risk premium itself is also subject to seizure during the game (i.e., if the amount of the potential loss is also under the control of the gunfighter), there is no risk premium great enough to induce the tenderfoot to play because the tenderfoot can never hope to do anything but lose all assets brought to the table.[96] The game never takes place.

We do not intend to imply that regulators are amoral gunfighters who are eager to change the rules in unfair ways to benefit ratepayers at investors' expense. The hypothetical example only illustrates the serious problem that the power to change the amount of the loss or to change the rules retroactively (now conferred by the Court for slight losses) creates for calculation of an *ex ante* fair risk premium. And without a fair *ex ante* risk premium, to return to the earlier metaphor, the tenderfoot will not play (or at least will stop bringing new amounts to the game).[97] Thus, exercising the freedom granted by the Court in *Duquesne* will eventually force regulators to face the problem of an appropriate regulatory risk premium.

---

[96]   Note that this is analogous to embodying the regulatory risk premium in the AFUDC rate: if the plant never goes into rate base, the risk premium never has to be paid. Yet as noted above, granting the risk premium in cash seems contrary to the spirit of a used and useful rule. We see no ready way around this difficulty.

[97]   If the gunfighter joins the game in midstream, the tenderfoot will already have cash on the table. The gunfighter may be able to force the tenderfoot to contribute new amounts of cash by a threat, implicit or explicit, to expropriate what is already in the game. But sooner or later the tenderfoot is better off forfeiting what is already on the table than contributing anything new.

Yet explicit recognition of regulatory risk with payments of premia for the right to change the rules can impose destabilizing incentives also. Unless regulators perceive meaningful costs of regulatory change and voluntarily refrain from changing the rules and *unless investors believe they will refrain,*[98] the prospect of perhaps limitless[99] regulatory risk to investors forecloses the game entirely. But naming that risk and explicitly compensating for it may make regulatory change inevitable, if only because regulators may feel bound to exercise the right once ratepayers have paid for it.[100]

Thus, the *Duquesne* opinion creates a difficult problem for regulators and investors alike.

Definition of 'Slight'

The final issue concerning retroactive changes in rate base methodology involves the measurement of 'slight.' The Court recognized that an uncompensated retroactive switch from the modified prudent investment standard to the used and useful standard unambiguously imposed real losses on investors. However, within an imprecisely defined limit of constitutional magnitude these losses can legally be imposed.

---

[98] The problem is even worse: under *Duquesne*, today's regulators cannot completely bind future regulators, so investors have to believe current *and future* regulators will refrain from destabilizing changes.

[99] The Court limited risks from rule changes to losses it found to be slight. But as shown below, the losses were larger than it recognized. Moreover, it did not expressly limit the number of rule changes that impose such slight losses.

[100] To return to the junk bond analogy, this is equivalent to regulators' forcing a utility to default on its debt *because* ratepayers have paid a default premium in the interest rate. Such a policy would rapidly close the market for new debt to that company. Louis Kaplow (1986, 509, 513) argues that there is a "close kinship between uncertainty regarding government policy and market uncertainty" and thus, "as an initial hypothesis governmental transitions warrant the same treatment as market transitions: no transitional relief." Whatever validity these conclusions may have for unregulated industries, this chapter has identified clear moral hazard problems for regulated industries.

In stating that it had reaffirmed both *Hope* and the Pennsylvania decision, the Court obviously believed that the problem of compensating investors for exposure to regulatory risks arising from switches in methodology could be confined to reasonable levels and assumed to be slight in most cases.

The economic foundations of this belief are questionable at three levels. First, it is not clear at what level constitutional magnitude ought to be measured. The Court stated that the canceled plants involved a $35 million investment for Duquesne, or 1.9 percent of its total rate base. Yet "more than 100 nuclear power plants worth $30 billion have been abandoned in recent years" (Greenhouse 1989, A1). The Court's decision to examine the definition of slight on a company-specific basis leaves open the cumulative impact industrywide. For example, given that cases are often settled without ultimate legal rulings, the *Duquesne* opinion may influence future settlements of disputed regulatory rules in a way that amounts to a large cumulative loss for the nation's regulated investors without an offsetting chance for a gain, even if the losses to Duquesne Light were indeed slight.

Second, the Court did not limit the *number* of 'slight' losses that could be imposed. Many small losses could add up to a large one in a hurry, especially given the third problem.

The third difficulty is that the actual losses were larger than the Court recognized. The Court, undoubtedly relying on the evidence before it, based its decision on the figures cited at the outset of this chapter: the loss was a 1.9 percent reduction of its $1.8 billion rate base and a 0.4 percentage point reduction in Duquesne Light's annual allowed rate of return.[101] Both measures are flawed.

---

[101] Earlier we assumed the Court's test referred to outcomes under traditional, symmetric regulation. The logic of the Court's reasoning might instead have called for a comparison of the end result of the switch with the range of possible variation of the prior ratemaking methodology. In the case of *Duquesne* this would have been the modified prudent investment standard, not the prudent investment standard. Thus it is not quite clear whether the test should be with the range of any acceptable methodology, against the prior methodology, or against the least risky methodology. Figures 2-8 and 2-9 illustrate that the window of constitutional magnitude could be extremely large if it is defined in terms of possible outcomes under extremely risky

The rate base represents the assets backed by debt, preferred equity, and common equity. But common shareholders bear the loss: the bondholders and preferred shareholders expect to be paid out of Duquesne Light's general revenues regardless of the regulatory treatment of the CAPCO investment.[102] Thus, the more appropriate test is the loss as a fraction of the *common equity* rate base. *Value Line* reports that Duquesne's common equity ratio was 34.5 percent in 1981 and 36.8 percent in 1982.[103] If we take the latter figure, the loss was 5.3 percent [$35 million/($1,800 million x 0.367)] of the amount put up by those who actually bore the loss, nearly three times the proportion recognized by the Court.[104]

The other measure used by the Court also has economic infirmities. The loss affected not just the first year's rates, but rates for as long as the amortization of the CAPCO plants' cost would have been authorized by the PUC's initial ruling. If the loss is to be calculated as the decrease in

---

prior ratemaking methodologies. Or perhaps "slight" must be measured as a comparison between the expected values under the old and new rules. Another interpretation is that it is defined in terms of the range of error in estimation of the cost of capital in the prior ratemaking methodology. These problems will obviously have to be addressed in any future cases involving the definition of constitutional magnitude.

[102] This statement embodies the assumption that either the debt used to finance Duquesne's CAPCO investment was not backed solely by the CAPCO assets, or if it were, Duquesne would find it more expensive to default on this debt (because of higher cost for future debt) than to redeem it. We believe the first assumption is quite likely; there certainly is no evidence in the opinion or the briefs of the parties to the contrary.

[103] *Value Line Investment Survey*, Duquesne Light, June 23, 1989.

[104] Recall the theorem (note 47, above) that rate base equals value to investors if the expected rate of return equals the cost of capital. If that condition were satisfied despite the rule changes in process, the $35 million loss in rate base would measure the lost value to investors between the pure prudent investment standard and the used and useful standard. If the Court is correct that Duquesne Light was on a compensated version of the modified prudent investment standard, the loss would equal the present value of the return of capital authorized in the original PUC decision. That would be less than the loss in the rate base if the conditions of the theorem were otherwise satisfied under those rules.

a single year's rate, the present value of the future lost amounts should be calculated and treated as the loss this year.[105] That procedure, if implemented correctly, at least provides an economically valid measure of the loss at issue in the case.[106]

Thus, if *Duquesne*'s method of calculating the loss is taken by courts and regulators as precedent, rather large property takings may be dismissed as of less than constitutional magnitude. If this is the case, *Duquesne* may not turn out to be the reaffirmation of the *Hope* standards that the Court apparently intended. This is true both because of the Court's definition of slight and because it is a much harder task to compensate investors for the risks they bear as a result of possible adverse changes in the rules of the game than it is to compensate them for the risks inherent in a consistent application of the initial rules.

REGULATORY RESPONSES TO DUQUESNE

The *Duquesne* opinion states "today we reaffirm these teachings of *Hope Natural Gas.* . . ." 109 S.Ct. 617. While the Court intended to reaffirm *Hope*, the economic logic of the *Duquesne* opinion imposes new requirements on regulators seeking investment in high-risk areas such as new base load electric generation.

---

[105] An analogy to the bias from looking at only one year's lost return is as follows. Suppose a customer adds $1,000 to a $10,000 floating-rate certificate of deposit. The customer knows there is a range of possible interest rates that he or she might earn. *Duquesne* says it therefore is constitutionally acceptable after the customer has added the money for the bank to change the rules to take the new $1,000 as a previously unmentioned fee, as long as the unexpected loss doesn't result in a rate of return lower than what the customer might have ended up with under the original rules. And when the Court measures the loss, it looks not at the amount taken, $1,000, but only at the first year's lost interest on that amount, say $80. That lost interest is a small fraction of the amount deposited ($80/$11,000 = 0.7 percent), well within the range of uncertainty in interest rates, so the Court's ruling is that the taking does not reach constitutional magnitude.

[106] A switch from modified prudent investment to the used and useful test imposes additional risks to investors that cannot automatically be deemed slight. For example, compare the additional risk premia required when going from, say, a 50 to a 100 percent loss in Table 2-1.

Whether the Court also has reaffirmed rate regulation as it has evolved since *Hope* depends on whether investors and regulators can agree to a mutually acceptable regulatory contract that meets the new tests set forth in *Duquesne*. We suggest that this agreement may be quite hard to reach.

First, the Court's *presumption* that the allowed rate of return accounted for regulatory risk does not hold generally. Indeed, we can say flatly that regulatory risks are not captured adequately by traditional rate-making methods used in regulatory proceedings. These methods attempt to equate the allowed rate of return for the regulated firm to the cost of capital. The cost of capital by definition is the *expected* rate of return, while the allowed rate of return should equal the equity equivalent of the fair *promised (or target)* rate of return given the regulatory risk.

The gravamen of the opinion is that regulatory risks must be specifically identified and a risk premium added to the otherwise applicable cost of capital when establishing the allowed rate of return. Unfortunately, methods for estimating the effects of regulatory risk on the necessary allowed rate of return are inadequate because the probability and expected size of the loss are difficult to determine. Moreover, the fair rate of return may be so high that it triggers even more regulatory risk, exacerbating the problem.

The problem is even worse when the size and likelihood of a loss are influenced by regulatory decisions, as is ordinarily the case. In this case, investors may perceive potential losses to be based on subjective and political factors. Explicit payments to investors for the regulatory risk may increase the likelihood of investor losses because of moral hazard: once having paid for the right to impose a loss, regulators may feel free -- even bound -- to do so.

An important consequence of *Duquesne* is that retroactive changes in the 'rules of the game' become an inherent risk in regulation and will be deemed proper as long as (a) the regulatory commission adjusts, *ex ante*, the allowed rate of return to reflect the fact that the rules may change during the game, or (b) the losses from the change are slight. Investor perceptions of an increased risk of future regulatory change are inevitable under these conditions, particularly given the underestimated economic loss in *Duquesne*.

An economic environment with increasing business risk, combined with a perception of high regulatory risk, may cause serious problems, including underinvestment in regulated industries and economically inappropriate incentives for industry operation. The regulatory risk premium that regulators are willing and able to pay in these circumstances may be simply not enough to overcome these problems.[107]   In these circumstances, *Duquesne* may put the last nail in the coffin of the prudent original cost rate base/opportunity cost of capital standard that has evolved since *Hope*.

Unless regulatory institutions change to accommodate the economic realities identified above, failure to account explicitly for regulatory and other asymmetric risk will usher in a new era of an undercapitalized public utility sector. Regulated firms will have strong incentives to defer investment and utilize small scale technology that is below minimum efficient scale. We are already beginning to see a 'reverse Averch-Johnson-Wellisz (A-J-W) effect,' whereby some public utilities will be starved for capital.[108]   Potentially inefficient forms of industry structure may also emerge as regulated firms minimize the assets exposed to opportunistic behavior by regulators.

---

[107]   As noted above, one way around some of these problems is to offer compensation through some device other than the allowed rate of return. However, moral hazard remains if an insurance premium or similar fee is added to the approved cost of service, and *ad hoc* adjustments in other cost items may not be a sufficiently reliable remedy for investors.

[108]   The 'Averch-Johnson-Wellisz (A-J-W) effect' of alleged incentives for overinvestment in regulated firms depends crucially on the assumption that the rate of return allowed by investors exceeds the cost of capital (Averch and Johnson 1962; Wellisz 1963; Baumol and Klevorick 1970; and Bailey and Malone 1970). To the extent that the A-J-W effect has been inferred from an allowed rate of return in excess of the cost of capital, the entire literature must be re-examined in light of the above demonstration that such a relationship is not *per se* an incentive for overcapitalization. That is, an allowed rate of return above the cost of capital but below that required to compensate for regulatory risk leads to incentives for *under*capitalization.

We see four possible regulatory responses that may follow in the aftermath of *Duquesne*. The first possible regulatory response is to reaffirm *Hope* by meeting the requirements imposed by *Duquesne*. To do this regulators will have to announce the rules of the game ahead of time and account for regulatory risk explicitly.[109] This scenario has the best chance of success if regulators choose ratemaking mechanisms that minimize regulatory risk and refrain from exploiting the window of constitutional magnitude by avoiding retroactive changes in the rules that are adverse to investor interests. If this is the approach, the (1) prudent original cost test is preferred over the (2) modified prudent investment test, which in turn is preferred to the (3) used and useful test.[110]

If regulators nevertheless wish to reserve the right to switch the rules while reaffirming *Hope*, they must develop objective means of measuring the risks and implementing the Court's test that are free from moral hazard. (We do not currently see a mechanism that would do so.) When considering retroactive changes in the rules, regulators will have to identify the risk structure inherent in the current scheme, as the Court did in *Duquesne*. They will also have to consider the consequences for future capital investment. They may conclude that a consistent application of past rules, whatever their drawbacks, would offset the short term benefits from switching to another ratemaking methodology.

A lot of things will have to go right for regulatory institutions to return to anything like the world we used to know. 'Constitutional magnitude' may be too vague a test for investors, and the necessary allowed rate of return may be unacceptably large for regulatory commissions. The stated or unstated intention of many electric utilities to avoid new con-

---

[109] See, for example, Joseph P. Kalt, Henry Lee, and Herman B. Leonard (1987) for one suggestion for a more precise definition of the regulatory compact. See also Glenn Blackmon and Richard Zeckhauser (1992).

[110] This future would be facilitated if the Court had an opportunity to rule on a case where the windfall loss to investors from adverse regulatory transitions exceeded the level of constitutional magnitude, particularly if the tests used to measure the magnitude were better. But we are aware of no good candidate case for such a ruling at present.

struction of generating capacity[111] reflects a situation where perceived regulatory risks have driven the necessary allowed rate of return to levels that regulatory commissions find unacceptable.

Second, regulators and investors may attempt to impose contractual limitations on regulatory discretion to change the rules. For the electric utility industry, this is essentially the path of cogeneration and power production contracts. But if regulators can change the regulatory compact within the bounds of constitutional magnitude so easily, can they not do the same to a formal contract that is retroactively found to be unused or unuseful, or merely unfortunate?

The imposition of contractual limitations on regulatory discretion may be aided by the current reluctance to sink new capital into traditional generation. In particular, regulators' ability to change contractual arrangements will be reduced if the entity they regulate, the utility, has as few assets "on the table" as possible.[112] In any case, attorneys for utilities, regulatory agencies, and investors may find their skills tested in finding contractual terms mutually acceptable to all concerned.

A third regulatory response seems to be a strong possibility in the mind of the Court:

> [R]igid requirement of the prudent investment rule would . . . foreclose a return to some form of the fair value rule just as its practical problems may be diminishing. The emergent market for wholesale electric energy could

---

[111] See, for example, *Electric Utility Week* (1985, 3). Public Service Company of Indiana shareholders voted overwhelmingly to "minimize future capital investment for the purpose of constructing new generating plants," until the treatment of investment in new plants improves in Indiana, because "the investments of the company shareholder should not be unreasonably put at risk through large capital programs to meet [future] demand."

[112] Contracts can be bought out at shareholder expense only by companies with the resources to do so. Thus, the payout to investors -- instead of reinvestment -- of cash generated by the business, which puts the money beyond the reach of regulators, is an economically rational response to regulatory risk.

> provide a readily available basis for determining the
> value of utility assets.[113]

Such a fair value rule would restore regulatory symmetry in a different way from the pure prudent investor standard. Instead of a system that prevents large losses to offset the regulatory prevention of large gains, the Court describes a system that permits investors the chance of large gains to offset the risk of large losses.

With this third response, rate regulation will have come full circle back to the fair value standard that predated *Hope*. Only this time, asset value will not be based on the subjective engineering studies that Justice Brandeis found so troubling (see footnotes 10 and 32, above). Rather it would be based on market prices for electricity, presumably determined by reference to the transactions that are evolving under the second, contractual response. A willingness by regulators to accept whatever the market dictates, even when such rates are far higher than they would have been under the pure prudent investment rule with an original cost rate base, would be a necessary step toward preventing a "heads I win, tails you lose" regulatory environment under this third response.

The analogy to the contingency enhancement for contingency legal work provided by Judge Williams (1991, 159-163) is an original contribution that points to a fourth solution (see Chapter 3 for discussion). As he notes, the size of the contingency enhancement defines the size of the contingency bar, just as we say the size of the regulatory risk premium defines the willingness to invest in regulated industries under asymmetric risk. Just as the courts have rejected an "individualized approach to fee awards," perhaps regulators could do the same for regulatory risk premia. It would be a matter of policy as to what the appropriate level should be, and such a system would not avoid problems such as moral hazard. But we can be confident that the correct answer for the regulatory risk premium is not zero, the answer currently embodied in regulatory practice, any more than would be a legal system that asks attorneys to take cases on contingency with zero fee enhancement.

---

[113] *Duquesne*, 109 S.Ct. 620.

In practice, we are likely to experience a combination of these four responses. The first will call for new and improved methods to measure regulatory risk and its effect on the necessary allowed rate of return (or for other cost adjustments) and the development of improved ways to reduce regulatory risk. The second will call for economic and legal analysis to design contracts with appropriate, enforceable incentives for all. The third will require development of better approaches to elimination of regulatory risk and to measurement of the value of assets based on market prices. And the fourth would require serious attention to the problem of defining the correct level of risk and dealing with moral hazard.

Chapter 3
Regulatory Risk: Objections to the Theory

> It is not sufficient to say that there are certain self-evi-
> dent truths; it is more accurate to say that all truth is
> self-evident. Because truth is self-evident, the best
> service that one can render a truth is to state it so clearly
> that it can be comprehended; for a truth once compre-
> hended needs no argument in its support . . . . [A]s truth
> is self-evident, so error bears upon its face its own con-
> demnation. Error need only to be exposed to be over-
> thrown (Bryan 1906, xii).

INTRODUCTION

Chapter 2 explored the broad consequences of regulatory policies that
impose the risk of a substantial loss without an offsetting opportunity for
a substantial gain ("asymmetric regulatory risk"). The logic of that
chapter contended that the rate of return allowed by regulators must be
in excess of the cost of capital in the presence of asymmetric regulatory
risk, such as the prospect of large cost disallowances.

It is common to assert, however, that utility investors are already compensated in the allowed rate of return for the risk of large disallowances,
such as arise for investments found imprudent or not 'used and useful.'
These assertions raise two questions: (1) could cost of capital studies in
theory be done correctly to compensate investors in regulated industries
for the risk of large disallowances even if current practice is wrong? and
(2) in practice, do cost of capital studies provide such compensation?
This chapter shows that infallible estimates of the cost of capital are sure
to provide downward-biased estimates of the necessary allowed rates of
return in the presence of such regulatory risks. It also shows that systematic errors in estimating the cost of capital cannot be expected to
correct for the bias. These conclusions justify the common perception
that rate regulation as sometimes practiced in the 1980s cannot survive
in the long run.

Our diagnosis of the problem and the recommended solutions that follow
from it represent a fundamental reformulation of many widely accepted

financial and regulatory principles.[1] In response, various commentators[2] have objected to (1) the conceptual framework, (2) our diagnosis of specific factual situations when applying the conceptual framework, and (3) the policy conclusions that we draw from it. Our response varies by type of objection.

To those who object to the conceptual framework itself, we seek to demonstrate that much of the disagreement stems from a misunderstanding of what we are saying. Consistent with the quote above from William Jennings Bryan, we hope here to restate the theory so that its common-sense logic is compelling to a wider audience. Once what we are saying and not saying has been made clear, we hope to achieve consensus that the theory in its simplest form is simply an irrefutable consequence of the law of averages applied to widely accepted definitions and is not unique to the kinds of regulatory risk that motivated our initial findings.

Responding to critics of the first variety can be frustrating, because the response to our central point often goes as follows. A skeptic says, "Kolbe and Tye (1991) say X." The skeptic then prepares an extensive rebuttal of 'X.' In most such cases, however, we did not say 'X.' At a minimum, we hope to clarify confusions of this sort in the pages that follow.

To the second group who accept the theory but object to our applications, we will illustrate these principles from time to time with what we hope is a persuasive view of the application of these principles to regulated industries in the recent past. Evaluations of the meanings of these

---

[1]   In addition to our chiefly economic analysis, there have been a large number of legal analyses of the constitutional issues surrounding such debates, particularly recently (Drobak 1985; Goldsmith 1989, 241; Pond 1989, 1; Darr 1990, 53; and Pond 1991, 111-116).

[2]   In connection with the original paper were the responses of two discussants (Williams 1991; Buchmann 1991). Two additional papers have been published by the *Energy Law Journal* (Cudahy 1991; Whittaker 1991). In addition, we have received a large number of private responses, both written and verbal. We quote liberally from these to give the reader a flavor of the response, but in many cases preserve the authors' anonymity.

facts will likely continue, even in the presence of a common set of principles; and the facts will vary from situation to situation. Honest differences of opinion are thus likely to continue when the theory is applied.

Lastly, to those who object to our proposed solutions, we readily acknowledge that differences over appropriate policy can continue even if there is agreement on the relevant model and facts. We welcome suggestions for improved ways of addressing these problems, especially in situations where we would be the first to agree that there are no clear answers.

A RESTATEMENT OF THE BASIC PROBLEM: ASYMMETRIC RISKS

Disallowance of prudently incurred costs creates an asymmetric payoff for investors: heads you lose, tails you just break even. Figure 2-4 illustrates the payoff in the presence of asymmetric regulatory risk. A basic theorem of rate regulation is that if the rate of return investors expect on the amount invested equals the *cost of capital*, defined as the expected rate of return in capital markets on alternative investments of equivalent risk, the market value of the investment today will equal the amount invested. (This result follows directly from the standard present value formula, used in Figure 2-4.) In Figure 2-4, the allowed rate of return on assets in the rate base is set equal to the cost of capital if there is no regulatory disallowance, but investors expect substantially less if there is such a disallowance. If there is any chance of a disallowance, then investors expect less than the cost of capital and the value of the investment is less than the amount invested.[3]

---

[3]   Of course, actual regulation does not work so precisely. Regulators can never be sure the allowed rate of return is exactly equal to the cost of capital, and chance events mean investors can expect realized returns above or below the cost of capital even if regulators have estimated everything perfectly. To focus the discussion, here we define regulatory risk as the chance of a *large* cost disallowance, one too big to be encompassed by the ordinary 'noise' in utility returns, that is *not* offset by a corresponding chance for an equally large gain. Chapter 2 discusses these issues in detail.

The fundamental intuition is quite simple: if a utility is promised the opportunity to earn, say, 15 percent on equity if things go as planned, and if it may earn much less than that if things go poorly, the rate of return truly expected is below the 15 percent opportunity that was promised. Therefore, the value of new investment is below its cost. Investors will not willingly supply capital under such conditions.

In the presence of asymmetric regulatory risk, fair treatment of investors requires that regulators must set the allowed rate of return on assets that do get into the rate base at a level higher than the cost of capital for the overall expected rate of return to equal the cost of capital.

The points made so far are discussed in considerably more detail in Chapter 2. What we are saying is in principle quite simple. Indeed, once the essential point is grasped as a fundamental feature of the law of averages, the conclusion is irrefutable as a matter of logic. Suppose that in some future states of the world ('bad outcomes') investors expect to lose large sums of money. In that case, investors must expect to earn considerably more than the cost of capital when losses do not arise (when 'good outcomes' occur) if they are to expect to earn their cost of capital overall. The implication is clear. Any regulatory system that incorporates a nonzero probability of a downside disaster because the investment is deemed to be imprudent or unused or unuseful will be fundamentally biased if the investor expects to earn only the cost of capital for used and useful, prudent investment. That result follows inexorably from the law of averages.

CONFUSION OVER 'THE COST OF CAPITAL' AND 'EXPECTED RATE OF RETURN'

At the outset, it is important to recognize that much of the objection to the conceptual framework itself stems from a confusion over terminology. Financial experts mean something very precise by 'the cost of capital' and the 'expected rate of return,' while these terms may have other connotations to experts in other disciplines. Once that confusion over language is cleared up, the principal conclusion of the conceptual framework may be seen as an intuitively obvious point about the law of averages and the relevant definitions.

The correct definition of the opportunity cost of capital should not be in dispute, as virtually all authorities define it as the *expected* rate of return in capital markets on alternative investments of equivalent risk (Brealey and Myers 1988, Chapter 7; Kolbe, Read and Hall 1984, Chapter 2). In this context, 'expected' is used in the statistical sense. As with all averages, the probability-weighted outcomes on the upside of the expected value exactly offset the probability-weighted outcomes on the downside. In the language of financial analysis, 'expected rate of return' most definitely does not refer to a 'hoped for' rate of return or a 'threshold' rate of return, as these terms are sometimes used in investment analyses. Nor does it mean the 'mostly likely' outcome.

The confusion arises if one applies common dictionary meanings of 'expected': "to consider probable or certain" or "to consider reasonable, due, or necessary" (*Webster's* 1976, 402). These meanings are altogether different from the statistical meaning of the cost of capital as an *expected* (or average) return. Hereafter, we use 'expected' only in the statistical sense:

> . . . the idea of expectation of a random variable is closely connected with the origin of statistics in games of chance. Gamblers were interested in how much they could "expect" to win in the long run in a game, and in how much they should wager in certain games if the game was to be "fair." Thus, expected value originally meant the expected long-run winnings (or losings) over repeated play; this term has been retained in mathematical statistics to mean the long-run average value for any random variable over an indefinite number of samplings. This holds whether a large number of samplings will actually be conducted or whether the situation is a one-trial affair and we consider hypothetical repetitions of the situation. Over a long series of trials, we can "expect" to observe the expected value. At any *single* trial, we in general cannot "expect" the expected value; usually the expected value is not even a possible value of the random variable for any single trial . . . (Hays and Winkler 1970, 136-137).

It is now widely accepted that regulators should seek to establish rates such that investors can *expect* to earn their cost of capital on their investments (or at least on their prudently incurred, 'used and useful' investments).[4] The logic is simple: if the *expected* rate of return on investment in the regulated industry is equal to the *expected* rate of return on alternative investment of equivalent risk -- i.e., the cost of capital -- the value of the cash flows investors *expect* will equal the amount they invest.

ILLUSTRATIONS OF REGULATORY ASYMMETRY

In practice, the widely accepted axiom that investors should expect a rate of return equal to the cost of capital is implemented when regulators equate the allowed rate of return on investments *in the rate base* with their best estimate of the cost of capital. However, this widely accepted axiom is not implemented in practice so as to achieve the stated intent: mere equation of the allowed rate of return on investments in the rate

---

[4]   Stewart C. Myers (1972) is the basic source of this regulatory theorem. It is also discussed in Kolbe and Read, with Hall (1984, Chapter 2). The use of the cost of capital so defined as the basic standard in rate regulation is consistent with the basic legal decisions on the topic, *Bluefield Waterworks & Improvement Co.* v. *Public Service Commission*, 262 U.S. 678 (1923) ('*Bluefield*') and *Federal Power Commission et al.* v. *Hope Natural Gas Co.*, 320 U.S. 591 (1944) ('*Hope*'). *Bluefield* held that a regulated firm:

> . . . is entitled to such rates as will permit it to earn a return on the value of the property which it employs . . . equal to that generally being made . . . on investments in other business undertakings which are attended by corresponding risks and uncertainties. 262 U.S. 678, 692.

Hope held that:

> the return to the equity owner should be commensurate with returns in other enterprises having corresponding risks. That return, moreover, should be sufficient to insure confidence in the financial integrity of the enterprise, so as to maintain its credit and attract capital. 320 U.S. 591, 603.

From an economic perspective, these criteria are met if the rate of return equity owners expect is the cost of capital.

base with the cost of capital need not, and in recent years *has* not, permitted regulated investors to expect to earn their cost of capital.
Moreover, this problem is due not to any defect in the way the cost of
capital is estimated nor to a systematic bias in estimation of the other
costs of service. Rather it arises from a fundamental flaw in the
prescription to equate the allowed rate of return to the cost of capital.

Suppose that regulators allow a rate of return on *assets in the rate base*
equal to the cost of capital but reserve the right to disallow capital
outlays found to be 'imprudent,' 'unused,' or 'unuseful.' As shown in
Figure 3-1, investors therefore *expect*[5] some of the total investment to
be declared imprudent and some to be declared unused and unuseful.
Under such a scheme investors will expect the rate of return on the total
amount they invest to be less than the expected rate of return on assets
in the rate base. The allowed rate of return on the total investment,
therefore, will be less than the cost of capital applied to the total
investment, as a matter of simple arithmetic, as long as there is a
nonzero probability of a disallowance.

The logic of Figure 3-1 demonstrates the need, whenever there is a
downside risk of disallowance, for a fundamental distinction between the
allowed rate of return and the *cost of capital*, which are often treated as
synonymous. The allowed rate of return on used and useful, prudently
incurred investment in the rate case cannot be equated to the cost of
capital if there is *regulatory asymmetry*.

Despite this logic, ratepayer interests often have denied that any
adjustment to the allowed rate of return should be made to account for
regulatory risk. In support of this view, they assert that (1) all examples
of regulatory risk already have been accounted for in calculating the
opportunity cost of capital. Alternatively, they assert that (2) the risks
facing regulated firms are no different from the class of risks facing
unregulated firms. Accordingly, they conclude that regulators can safely
ignore regulatory risk when deciding the allowed rate of return and
choosing whether to impose regulatory losses on investors.

---

[5]   Once again, the reader is reminded that this term is used throughout in the
probabilistic or statistical sense. After the investments are in place, of course,
realized disallowances will not ordinarily equal expected disallowances.

Figure 3-1
Diagrammatic Illustration of the Basic Point

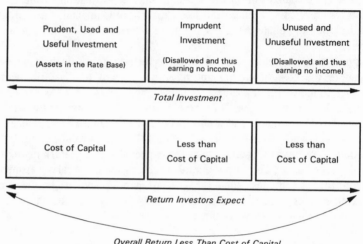

As discussed in Chapter 2, the Supreme Court's recent opinion in *Duquesne Light Co., et al.* v. *Barasch, et al.*, 109 S.Ct. 609 (1989) addressed some of the regulatory dilemmas that have resulted from these bad outcomes. The Court in *Duquesne* affirmed that regulatory risks must be accounted for in establishing the allowed rate of return. But in view of the Court's presumption that the allowed rate of return already compensated for the pre-switch level of regulatory risk, it is vital that one fully understand the level of regulatory risk actually embodied in allowed rates of return. The issues must be clearly understood as part of any requirement to conform future regulatory practice to the economic principles of *Duquesne*.[6] Among the objections to our logic we focus specifically on claims that the allowed rate of return on utility equity as

---

6   When issued, the *Duquesne* decision was widely regarded as giving a green light to a permissive standard for regulatory disallowances. The thesis of Chapter 2 was that *Duquesne* was not the victory for ratepayer interests that many believed. At least one article with apparent sympathies for ratepayer interests would agree (Copeland and Nixon 1991).

conventionally determined automatically compensates for such risks. We conclude it does not; in fact, the allowed rate of return may well not even be *close* to the level needed for fair compensation in the face of such risks.

This means regulation as sometimes practiced in the 1980s cannot survive. If regulators choose to retain policies that impose such risks, either new procedures will be required to determine a fair (and much higher) allowed rate of return (or to offer other compensation approaches discussed below), or new investment subject to such risks will be difficult or impossible to finance.

THE JUNK BOND ANALOGY ONCE AGAIN

Chapter 2 started with an analogy of regulatory disallowances and junk bonds. We chose the vehicle of junk bonds for illustrating the appropriate financial concepts because the distinction between a *promised* and *expected* rate of return for junk bonds was neither original with us nor seriously in dispute among financial experts. Indeed, disappointed junk bond investors were being reminded daily of the key distinction as the original article was being published (Kolbe and Tye 1991). Thus, we hoped to launch our discussion of regulatory risk from a point of departure where there was little disagreement.

All analogies end at some point, and unfortunately some critics have lost sight of our essential point by digressing into an attempt to rebut propositions unrelated to our central themes. They either (1) offer a critique of standard finance textbook analysis of junk bonds, (2) accuse us of trying to give stockholders the same downside protection against risk as bondholders, or (3) point out that stockholders generally have greater upside potential than do bondholders.

We address these objections to the junk bond analogy only briefly here because they represent digressions from the main points we are trying to make. First the distinction between the *promised* rate of return and *expected* rate of return for junk bonds is standard fare in introductory corporate finance textbooks, to which we refer any readers with remaining doubts (Brealey and Myers 1988, 561-562).

With regard to the second objection, our point was simply that the analytical mechanism for computing a regulatory risk premium is similar to the analytical mechanism for computation of the default premium for junk bonds. At that point the analogy ends. Indeed, the same mechanism would apply throughout the spectrum of investments from treasury bills (zero default risk) to penny stocks whenever default risk exists. We are not attempting to give utility stockholders the same protection against risk as enjoyed by bondholders. Even after the effects of asymmetry have been corrected for both a company's stocks and bonds, the utility's stock will still have greater risk than its bonds (more upside and downside potential) and a higher cost of capital.

Lastly, our critics misstate our analysis to impute us as saying that utility stock investors can expect no possibilities of upside gains over the cost of capital. We are saying no such thing, as a cursory examination of Figures 2-5 and 2-6 shows. We are well aware that utility shareholders sometimes do better than expected, due to regulatory lag, for example. A fair reading of Chapter 2 shows that we assume that bond returns are truncated at the promised rate of return (Figures 2-2 and 2-3) but this is not true for stocks (Figures 2-5 and 2-6). So this objection attempts to rebut something we never said.

In summary, the junk bond example was offered as what we hoped would be an uncontested example of the mechanics of how risk premia are computed in competitive capital markets for any security. For many with training in financial economics, the analogy has proved helpful in understanding regulatory risk. To those who find themselves bogged down by irrelevant objections to the junk bond analogy, we urge them simply to move on to a case of asymmetric regulatory risk, where the basic proposition can be derived as a simple consequence of the law of averages applied to the relevant definitions.

To those who have trouble with the junk bond analogy, we offer another, perhaps one which will evoke more empathy from lawyers who try rate cases but who are unfamiliar with finance theory. Suppose one wished to set reasonable fees for successful plaintiffs' attorneys who served on contingency. Since there would be no fee if the case is lost, the hourly fee must exceed the usual fee (the 'lodestar fee') by an amount dependent on the probability of losing.[7] This risk premium is referred to as the 'contingency enhancement.'

One could impose the contingent fee issue on Figure 2-4 where (1) the probability of default or disallowance is replaced by the probability of zero fee, (2) the 'regulatory risk premium' is replaced by the 'contingency fee enhancement,' (3) the 'promised rate of return' is the 'contingency fee' (sum of the 'contingency enhancement' plus the 'lodestar fee') and the 'cost of capital' is replaced by the 'lodestar fee.'

The key to the analogy, of course, is that the very definition of the 'cost of capital' is equivalent in concept to the 'lodestar fee,' if ". . . a large law office, at least, can diversify the risks over a large range of cases,"[8] producing an average contingency fee equal to the lodestar fee. It is exactly such an average that the opportunity cost of capital seeks to measure in concept.

Suppose one attempts to measure the appropriate contingency enhancement by collecting a large set of data on average (or 'expected' in the statistical sense) fees for comparable work (read 'the cost of capital' in financial studies in rate proceedings). If the sample were data on fees for work not done on contingency, clearly the result is downward-biased. *But even if the average were for all legal work done on contingency* (the sample contained exactly the same risks of zero payment for unsuccessful suits as for the case at hand), *the average of such a population would be downward-biased* (because what is really needed is the average fee *winners* get).

Anticipating some of the discussion below, the analogy may be pursued one step further. Many of the objections to our central thesis are premised on a belief that cost-of-capital studies properly done will automatically include the proper risk premium. These objections to the principles in Chapter 2 are equivalent to saying that a study of comparable lodestar fees could fortuitously produce the correct number for the

---

[7]   See *King* v. *Palmer*, 906 F.2d 762, 769-770 (D.C. Cir. 1990).

[8]   *Ibid.*

contingency fee and contingency enhancement, if done in a sufficiently biased way. Only if the study of average returns for all contingency work (both successful and unsuccessful) were systematically biased upward, to the point where it fortuitously gave the average winner's fee, would it produce the correct answer for the appropriate contingency fee for the 'successful outcome.' We show below that this fortuitous result is not likely.

An editor of an earlier article has suggested the analogy of regulatory risk to investing in oil wells. Suppose that 80 percent of drilling investments are dry holes. The accountant has (at least) two choices for the denominator when summarizing profit information. One is to include the investment on all wells, successful and dry. This would correspond to the *ex post facto* return on investment actually made. Or the accountant could have written off the dry holes and published the return on successful wells, producing a rate about five times the other. Both would be 'correct' summaries of information. So long as the oil company's pricing decision does not have to be justified by appealing to a particular profit number, it does not really matter how the information is summarized.

However, there is a big difference between regulated firms and unregulated firms in reporting profit information. The regulated firm makes its investment decisions and then must *justify* the prices it wants to set in terms of a reasonable rate of return. The element of justification makes the big difference. The regulated firm's compensation ceiling will be calculated by multiplying two numbers together. The numbers for rate of return and investment have to be determined in a consistent manner.

A CRITICISM: IS FIGURE 2-4 RIGHT?

The logic of these concepts has received two diametrically opposed responses. One response is that the propositions are so patently obvious as to hardly bear repeating.[9] According to this mode of thinking any regulatory regime that routinely exposes investors to large risks of massive losses without additional compensation is clearly not providing a fair return. Regardless of what finance theory may be perceived by some to tell us the risk, investors simply will not be attracted to industries where they can routinely expect to incur large losses through regulatory disallowances with no equivalent hope for offsetting gains on the upside.

The other, diametrically opposed, response is that the above arguments are nonsense and are refuted by both the Capital Asset Pricing Model (CAPM) and Discounted Cash Flow[10] (DCF) methods of measuring the

---

[9]   One of our anonymous referees from an earlier draft of this chapter observed:

> This paper makes a simple, but useful point . . . . Presenting it to an audience of economists might be a little like preaching to the converted. However, the paper does point out the implications of some basic financial principles that, while simple, can easily be forgotten.

[10] 'Discounted Cash Flow' is a term of art in cost of capital estimation. Ordinarily, 'discounted cash flow' and 'present value' or 'present worth' are synonyms: the result of a calculation designed to express the value today of a stream of (possibly uncertain) future cash flows. In cost of capital estimation, however, the 'DCF method' attempts to infer the discount rate (sometimes called the 'DCF rate of return' or 'internal rate of return') that makes the value of future expected cash flows to utility investors equal to the price of the utility's stock. The most basic 'DCF formula' for the cost of capital is:

$$k \quad = \quad (D/P) \ + \ g$$

where k is the cost of capital, D is the dividend expected at the end of the next period, P is the stock price, and g is an assumed perpetual, steady growth rate for dividends. This DCF cost of capital estimation formula is derived from a simplified present value formula (or DCF formula in the standard sense), which is:

$$P \quad = \quad D/(k - g)$$

This simplified present value formula holds when the company has no valuable growth options and investors expect dividends to grow at a steady rate forever. See, for example, Kolbe, Read, and Hall (1984, Chapter 3) for more details.

In this chapter, we use 'DCF method' to refer to the cost of capital estimation technique, not the present value formula.

cost of capital.[11] These skeptics often claim that all risks, including regulatory disallowances of costs deemed to be imprudent or 'unused and unuseful,' have been accounted for in the measurement of the cost of capital, or they rely on the logic of CAPM to claim that any risks not included in the cost of capital are fully diversifiable and need not be accounted for in any event.

When we have probed the second group for the reason critics believe we are wrong, most of them argue that Figure 2-4 is incorrect. Instead, they believe that traditional methods for estimating the cost of capital produce an allowed rate of return on the 'no disallowance' branch of Figure 2-4 that is something akin to the promised rate of return on a bond, so that the utility's expected rate of return on equity equals the cost of capital even when asymmetric regulatory risk is present. Others argue that since all of the facts are known to investors before the regulatory risk is resolved, they will be fairly treated even if Figure 2-4 is right.

Response to these objections goes easiest by explaining first what we are not saying. We are not saying that CAPM and DCF are *necessarily* biased estimates of the *cost of capital*, nor are we rejecting any of the results that flow from these methodologies, given their assumptions. In fact, the more infallible are these methods in accomplishing their stated intent, the more certain one can be that the estimated cost of capital will provide a downward-biased estimate of the necessary allowed rate of return. Nor are we necessarily saying that all regulatory risks such as disallowances of high cost plants or other regrettable decisions are not diversifiable, at least in theory. Finally, we do not rely on market imperfections or investor misperceptions to reach our conclusions.

The short answer to those who believe the allowed rate of return routinely includes compensation for asymmetric regulatory risks is to quote the textbook definition of the cost of capital: the *expected* rate of return for comparable risk investments. All of the commonly used methods of estimating the cost of equity capital aim at the expected rate of return. If the analyst applies one of those methods correctly and if the

---

[11]   See discussion below for specific citations.

commission in question adopts that rate of return, the end result will be exactly as depicted in Figure 2-4. The law of averages implies that the weighted average of the cost of capital and something less than the cost of capital (i.e., a disallowance) *must* be less than the cost of capital if any weight exists for the 'disallowance' branch.

SPECIFIC OBJECTIONS TO OUR ANALYSIS

Some readers, however, find our short answer unpersuasive, despite the fact that it relies on an elementary property of the law of averages as applied to definitions of the terms that are uncontested in standard textbooks on the subject. These objections relate to (1) the conceptual framework itself, or (2) to the application of the results to particular historical events, or (3) specific policy recommendations that we and others have offered.

Objections to the Conceptual Framework

To the extent that the objections to the concepts themselves can be organized, they seem to fall into five classes:

1. Explicit accounting for regulatory and other asymmetric risk would shield investors in regulated industries from the risks common to unregulated industries.[12]

2. The Discounted Cash Flow (DCF) method of measuring the cost of capital automatically

---

[12]   As one critic puts it:
    . . . I am struck with what seems to be the core assumption on your part -
    - that the investor should be virtually guaranteed to earn the "expected"
    rate of return. You say, in effect, that if there is a chance that the
    investor will earn less than his expected rate of return, then the regulators
    should allow him to earn the "promised" rate of return which is at some
    higher level.

incorporates all risks, including regulatory risks.[13]

3.   The Capital Asset Pricing Model (CAPM) denies that diversifiable risk needs to be accounted for in the cost of capital; regulatory and other asymmetric risk is fully diversifiable and thus can be safely ignored.

4.   Perfect capital markets permit one to ignore regulatory and other asymmetric risks, because they will be recognized by investors.[14]

---

[13]   In reality, this objection is part of a larger set of concerns. As expressed by one private communication, there is a need for:

> . . . a complete explanation for why the observed rate of return for equity will fall below the promised rate of return in a regulatory regime where the observed (expected) rate embodies full information . . . . So my question is not whether there ought to be a regulatory risk premium, but whether there isn't one already in a correctly-performed rate of return analysis.

We focus principally on the belief that the DCF model accomplishes such a result, because that is the most concrete form of this argument.

[14]   See Buchmann (1991, 169), for example, who expresses concern that the need for a regulatory risk premium:

> . . . requires a finding that the market *still* does not recognize the risk of disallowance, which requires a high degree of naivete on the part of the investors or their advisors.

> This raises another issue. *If* the market is efficient and *if* the investors assimilate all available facts, assumptions made by many analysts, it is difficult to understand why a regulatory risk premium, over and above a market-indicated cost of equity, exists at all. Given the history of the last decade of ratemaking, replete with disallowances for one cause or several, why has the market not taken this risk into account already? [Footnote omitted.]

This objection is thus closely tied to objection No. 2. It starts with a belief that any proper measure of the cost of capital that relied on market data would automatically

5.   The model ignores opportunities by regulated
       firms for gains in excess of the cost of capital.

Each of these will be addressed in detail, but they can be refuted quite
simply. Contrary to the first objection, the junk bond example (Chapter
2) illustrates that competitive capital markets for unregulated assets work

---

incorporate a proper risk premium for regulatory risk. According to this logic a
regulatory risk premium over the cost of capital must rely on either market
imperfections (investor naivete) or an explicit historical switch in risk-bearing which
was not expected by investors. To which, skeptics deny that investors are so
ignorant as to ignore the risk of a switch or to deny that it could, in fact, happen.
In any event, they claim that investors have been properly warned as to the future.
As another correspondent put it:

> . . . if the investor has "been around" a little, this belief in the impossibil-
> ity of a rule change would brand him a fool.

> Your argument . . . assumes that the cost of capital (presumably
> measurable in the market) is determined independently of the possible
> decisions of the regulator. This can occur simultaneously with voluntary
> investment only if investors believe that the *possible* decisions are
> *impossible*.

> Voluntary investment would not occur until the "allowed rate of return"
> had an appropriate level. The market, assuming risk neutral investors
> would use your method of calculating an appropriate "allowed rate of
> return" as its method of calculating a "cost of capital." Thus, the
> regulator, if he used market information, would set an "allowed rate of
> return" sufficiently high to attract the voluntary investment.

> Essentially, you argue that the investor has a complaint because the
> regulator told him the rules wouldn't change, and if the investor had
> known that the rules might change he would have insisted on a higher
> return. . . . Although your argument is logical given its premises, I
> don't think many would accept the premises as representative of the
> investment community.

> If investors had a belief that a rule change event had a probability greater
> than zero under the modified prudent investment methodology, then they
> *did* receive compensation for the risk of rule changes. Now that the rules
> have changed, investors will have revised expectations about the
> probability of a disallowance and will adjust their required returns
> accordingly.

in just the same way as proposed for regulated firms. The second and third objections are mutually inconsistent. The DCF objection claims that regulatory and other asymmetric risks are real but already accounted for, while the CAPM objections claim that they may be safely ignored because they are fully diversifiable! Both procedures, however, are presumably designed to provide an unbiased estimate of the same number, the cost of capital, producing a paradox, indeed.

The averaging logic of Figure 2-4 holds regardless of how the cost of capital was measured and in no way contradicts the assumptions of any model for computing an unbiased estimate of the cost of capital. This result is most obvious for the CAPM methodology, since even our critics agree that nothing necessarily happens to the CAPM estimate of the cost of capital after newly exposing investors to a major increase in regulatory risk.[15] (For this reason, CAPM practitioners who disagree with us tend to do so on the grounds that regulatory risk should be ignored in estimating the cost of capital.) Since the DCF methodology seeks an unbiased estimate of the same thing as the CAPM methodology, one would expect a DCF estimate done right to have the same properties, i.e., to ignore the regulatory risk. We address below whether some systematic upward bias in the DCF methodology could automatically provide the correct rate of return in the presence of regulatory risk.

Objection Number 4 is more difficult to address because it is often articulated in idiosyncratic prose. To the extent that it can be reformulated into the language of modern finance theory, it seems to rest on a faulty premise. The logic of Figure 2-4 assumes no market imperfections to reach its conclusions. Indeed, it relies on the same assumptions about capital markets as does almost every axiom of modern finance theory. However, the issue of whether we could have accumulated the existing capital stock in regulated industries if things were as bleak as we say is a legitimate question. As discussed below, our answer is that the odds have changed so that what was once a remote theoretical possibility is now a reality investors cannot ignore.

---

[15]   The cost of capital would remain unchanged if the regulatory risk were entirely diversifiable, i.e., not correlated with other economic events.

Objections to Applications of the Concepts

Our critics also take issue with our use of the model to diagnose specific historical events in the regulated industries:

6.   After all, investors are allowed a rate of return on a specified base of assets above the risk-free rate because they bear risk; they have recently had to bear some losses, but that is just bad luck, not unfair treatment.

7.   Some investors bought the stock at reduced prices after a change in the rules was announced, and it would be inequitable for them to revert to the original regulatory scheme or to be compensated for the additional risk.

8.   Investors are irrational and have no sense of the past; they will still expect to earn their cost of capital under asymmetric rules for future projects even after they've been burned in the past. Investments are still being funded in regulated industries, despite the change in rules.

9.   The theory contradicts the 'capture model' of regulation, whereby regulators protect the regulated industry.

Objections to Policy Recommendations Based on the Analysis

A number of commentators have expressed other concerns:

10.   Other changes in regulatory institutions (incentive ratemaking, more economically efficient ratemaking, etc.) have offset the increased degree of regulatory risk.

11. Recent investment disasters are wholly retrospective; managers, regulators, and investors have learned from these investment write-offs and the future looks much brighter.

12. It would be difficult or impossible to implement a regulatory system that explicitly recognized regulatory risk.[16]

Similarly, critics sometimes assert that a specific allowance for regulatory risk would take away the chief tool now used by regulators to motivate efficient performance, that is, denial of cost recovery. This objection is often made in connection with the above objection that recognition of regulatory risk would protect investors from risks routinely borne by unregulated firms.

But these difficulties in implementation are no reason to reject the theory itself. Rather the criticism should be directed to regulatory institutions that sought to impose significant risks onto shareholders with little or no compensation.

RESPONSES TO THE OBJECTIONS TO THE CONCEPTUAL FRAMEWORK

Does Accounting for Asymmetric Risk Insulate Investors from Risks Common to Unregulated Industries?

The formulation of the problem in Figure 2-4 is frequently accused of trying to give investors in regulated industries a "free ride" on risk at ratepayers' expense. After all, don't investors in unregulated industries wake up some mornings to read that their stock in a savings and loan company is now worthless, etc.? And aren't those losses in unregulated

---

[16] See Buchmann's (1991) discussion of "evidentiary obstacles" and "other difficulties," for example.

industries taken into account when measuring the opportunity cost of capital?

The simple answer is that the logic of Figure 2-4 is exactly the same as for unregulated securities markets. The object is to do no more and no less for investors in regulated industries, as the analogy to the default premium for junk bonds discussed in Chapter 2 shows.

Other analogies to the need for a regulatory risk premium abound in competitive capital markets. South African gold stocks generally sell at low price/earning ratios and pay higher dividends than the typical stock. The same is true for cigarette companies who are viewed as 'cash cows.' Why then don't investors snap them up, driving up their stock price?

The answer may well be that investors perceive some nontrivial chance that these stocks will incur events equivalent to the 'default' in Figure 2-4. In the first case it would result from civil disorder in South Africa. In the second case health claims by smokers may prevail in the courts, causing large losses to investors. These higher promised returns if these adverse outcomes do not occur are thus functionally equivalent to the default premiums for junk bonds.

The claim that we wish to 'insulate investors from risk' via the logic of Figure 2-4 is difficult to understand. No one claims that investors in junk bonds have been insulated from risk because promised rates of return are very high. We are merely proposing that the very same mechanisms for risk compensation must be used for regulated markets as for unregulated markets. This compensation mechanism is premised on the assumption that investors in regulated firms will in fact bear these risks, exactly contrary to the claims of our critics.[17]

Of course, the utility does not have a promise that the allowed rate of return will actually be earned absent a regulatory disallowance, only that the utility will have a *fair opportunity* to earn that rate of return. Some of our critics have incorrectly seized on the technical term 'promised' to

---

[17] For an illustration of mechanisms to achieve such results, see James A. Read, Jr. (1989, 25-30).

infer that the rate of return in the absence of default was guaranteed. Thus, perhaps 'target' rate of return is a better phrase than 'promised' rate of return when speaking of utility equity.

Does Use of the Discounted Cash Flow (DCF) Methodology Automatically Account for All Risks?

Those who subscribe to what we are calling the 'DCF objection' accept the idea that regulatory and other asymmetric risk must be accounted for but argue that the DCF methodology accounts for all such risks.[18] This objection often comes from those who have developed a high degree of faith in DCF as a technique for measuring the cost of capital.[19]

---

[18]   Recall that 'DCF methodology' here refers to the cost of capital estimation technique, not the present value formula. See above, footnote 10.

[19]   One skeptic put it as follows:

> The amount of risk premium that a particular security carries is directly related to its perceived risk. The "required" rate of return is a threshold rate that an investor must believe that he has an *opportunity* to earn before he will make a particular investment with its associated risks. This is the rate that is testified to by cost of capital witnesses.
>
> *   *   *
>
> Certainly the DCF method encompasses all of the risks that are known to or imagined by investors. Under your assumptions, the "expected" rate of return, being virtually guaranteed, should approach the risk-free rate of governmental securities. The "promised" rate of return would then equate to the results of the DCF calculation, providing a risk premium over the risk-free, or nearly risk-free "expected" rate of return.
>
> *   *   *
>
> Since the investor must be assumed to have a full knowledge of the investment in question, he therefore is aware of all risks *including* regulatory risk. Since investors are making daily investments in utility stocks, it would seem obvious that the returns of each company, and therefore the DCF returns, are equal to investors' required returns. In conclusion, I believe that the "promised" rate of return is the rate that is addressed by regulatory commissions and by cost of capital witnesses, not the "expected" rate.

Note that this criticism uses terms in a manner contrary to the textbook definitions cited above. We take this as a claim that such confusion could fortuitously produce

The claim that the need for a regulatory risk premium is incompatible with the DCF methodology can be laid to rest with a simple thought experiment. Let us return to Figure 2-4 and assume for the purpose of argument that the DCF measurement exercise has produced an absolutely infallible measure of the cost of capital. If so, then the law of averages tells us that the number produced by DCF for the cost of capital is a downward biased estimate of the necessary allowed rate of return if there is any nonzero probability of a large cost disallowance (or other asymmetric) outcome. *DCF measures the cost of capital, if done well, but the cost of capital is simply the wrong rate of return to allow.*

Despite this logic, the reasoning of the DCF objection seems to run something like this:

1.   The price of a utility's stock is supposed to reflect all information, including any asymmetric regulatory risk.

2.   Modern cost of capital estimation methods, such as the discounted cash flow (DCF) approach or the capital asset pricing model (CAPM), are based on analyses of the stock price. In particular, regulatory risk that affects the stock price will increase the measured dividend yield in the DCF model, at least, even if it does nothing to the CAPM.

3.   Therefore, whatever risks affect the utility must be incorporated in the cost of capital estimated by one of these methods. If the allowed rate of return equals such an estimate, in this view, investors will automatically be compensated for all risks.

---

the correct results for the promised (or target) rate of return via the DCF methodology in practice, despite the plain meanings of the definitions.

According to this logic, regulators automatically account for risk of regulatory disallowances when they set the allowed rate of return to an estimate of the cost of capital produced by the DCF methodology. Skeptics are saying in effect that future regulatory risks must be low because DCF estimates of the cost of capital are saying they are low.

We agree with the first two of these numbered statements, but the conclusion in the third does not logically follow from them. To explain why, we need to explore the short answer given above in more detail.

Returns on Bonds. As before, we start with the bond analogy. The yield to maturity on a corporate bond is calculated by a DCF method.[20] The terms of the bond spell out the timing and amounts of the cash flow that lenders will get if all goes well. In most cases, these cash flows consist of periodic coupon payments and eventual bond redemption at face value. The yield to maturity is just the discount rate that makes the present value of these promised future payments equal to the current *market* value of the bond. Thus it is the *promised (or target)* rate of return discussed above.

The same DCF technique could be used to infer the *expected* rate of return on the bond. In fact, the cost of capital of a bond is its expected rate of return too, not the promised yield that is ordinarily quoted. To obtain a DCF estimate of the expected rate of return on a bond, an analyst would need to forecast the cash flows lenders actually expect, which are the coupon payments and principal redemption *reduced by the chance of default.* The analyst then could calculate the discount rate that equates the forecast of the expected cash flows with the *same* market price of the bond. That discount rate is the bond's cost of capital.

---

[20] As a general definition, DCF techniques solve for the discount rate that equates the price of a security with the present value of the cash flows that accrue to that security's owners. The well-known 'DCF formula' for the cost of equity capital, $k = (D/P) + g$, is only a special case of a more general approach. See, e.g., Kolbe, Read and Hall (1984, Chapter 3) for discussion of DCF approaches to cost of capital estimation.

Thus, as shown in Figure 3-2, the bond's current market price can be thought of as resulting from *either* of *two* present value calculations: (1) the promised cash flows discounted at the promised rate of return (i.e., the yield to maturity), or (2) the expected cash flows discounted at the expected rate of return (i.e., the cost of capital). Both present value calculations, properly done, equal the bond's market price.

Figure 3-2
The Same Market Price Equals Two Means of
Calculating Present Values if There Is Asymmetric Risk

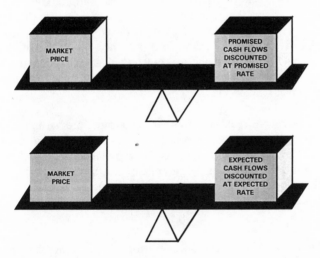

Note also that *both* of the initial conditions stated above are satisfied for these two rates of return: the bond price reflects all available information that matters to investors, and the bond's current market price is used directly in a DCF calculation to infer the rate of return. But there are *two* rates of return associated with that single bond price, not one.

Of course, it is the *promised* rate of return that has to be paid to bondholders, absent actual default, if bondholders are to be treated fairly. If regulators refused to permit utilities to make the promised bond payments, on the undisputed grounds that corporate bondholders actually expect something less because of default risk, bondholders would be shortchanged.

Returns on Stocks. We hope at this point that it is obvious that exactly the same arguments can be made for utility stocks. In the presence of asymmetric regulatory risks, it will be possible in principle to calculate *two* rates of return for the stock: the *'promised' (or target)* rate of return that the utility is to have an opportunity to earn with a good outcome, i.e., if the disallowance or other adverse event does not come to pass; and the *expected* rate of return, which reflects both the good and bad outcomes. Both of these, in principle, could be calculated using the same stock price, which reflects all risks known to investors. Both could be calculated with DCF techniques, one focused on the dividends investors could hope for if the bad outcome does not occur, the other focused on the dividends investors actually expect given that the bad outcome in fact may occur.

The expected rate of return on equity is the quantity ordinarily estimated by cost of capital analysts. This is clear from the theory and language of the DCF approach, which seeks *expected* dividend yields and growth rates. Yet the 'promised' or 'target' rate of return is what commissions must allow in the presence of asymmetric regulatory risk to give investors a fair opportunity to earn the cost of capital.

Loose Ends to the DCF Objection. Even readers who agree with us in principle at this point may still have some nagging doubts about whether we are right in practice. For example, given that the DCF model can be consistent with multiple rates of return, it is natural to ask whether some easy adjustment to the DCF method at least (if not the CAPM) would provide the sought-after fair target rate of return. Even better, might not some common mistake in the application of the usual DCF approach[21]

---

[21]   William Beranek and Keith M. Howe (1990, 192) state that:

> There is no evidence that DCF rate-case witnesses are aware of these principles [of consistency conditions with financial theory] let alone employ them in checking, tempering, and limiting their recommendations.

automatically yield the appropriate target rate of return, rather than the expected rate of return that is supposed to result?[22] Also, might not the regulatory risk raise the *expected* rate of return (i.e., the cost of capital)? If regulators allow this higher cost of capital, might not they automatically take care of the problem we've been discussing?

We address these loose ends in turn. It turns out that the picture is every bit as bleak as we have painted it.

DCF Estimates in Practice. We examine a simple case to evaluate the practical outcome of a DCF calculation. If the DCF method does not fortuitously yield the appropriate target rate of return in a simple case, it is unlikely to be able to do so in a more complicated one. Indeed, if we produce even one counterexample that shows the DCF model will not automatically produce the correct result, that should be the end of the matter: unless the DCF method in practice somehow gets the right target rate of return every time, the problems we raise here will have to be faced head on. Thus, the following counterexample should suffice to put the DCF objection to rest.

Assume as shown in Scenario 1 in Figure 3-3, that regulators observe a company with a book and market value of $100 per share and compute the cost of capital at 15 percent via the DCF method (a dividend of $10 is expected to grow at 5 percent forever). Thus, the $100 share value is expected a year hence to produce a dividend and share price jointly worth $115. Regulators suddenly discover that the utility has an

---

If the admittedly nonrandom sample of response to our initial article is any indication, it is fair to say that there is widespread disagreement, if not outright confusion, among practitioners of the DCF model. Given this lack of understanding of the fundamental principles inherent in the DCF methodology by those who profess to employ it, one cannot rule out *a priori* the possibility of the 'law of equal cheating' somehow producing the correct answer with the wrong application of a model designed expressly to estimate something else entirely.

[22] For example, investment analysts' forecasts provide *expected* growth rate estimates, which will tend to be pulled down by any regulatory risks embodied in the stock price. But other methods to estimate growth rates exist. Maybe one of these happens to yield the appropriate target rate of return.

environmental cleanup cost of $300 per share (far in excess of the book and market value of the stock) that will have to be paid off in one year. Regulators announce they will meet tomorrow to decide whether ratepayers or shareholders should bear the cost. Investors perceive that the decision could go either way, equivalent to a coin toss.[23] Furthermore, regulators restate their intention to equate the cost of capital to the allowed rate of return if they decide not to make investors bear the cost; as shown in Scenario 2 of Figure 3-3, they therefore ignore the new risk.

Figure 3-3

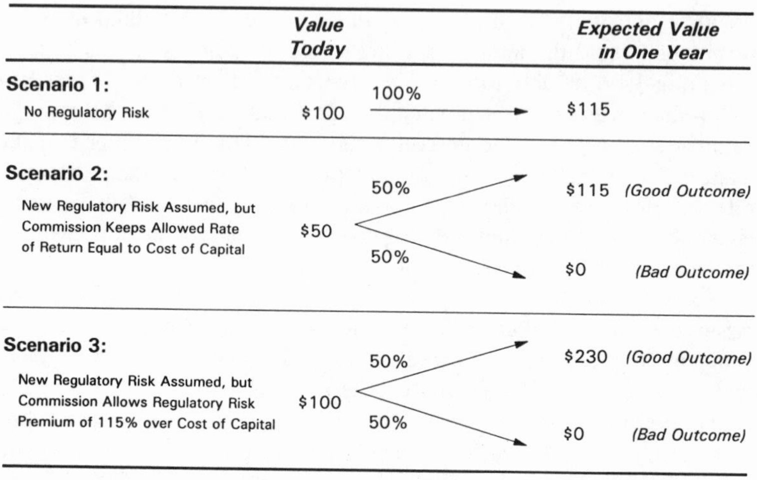

	Value Today		Expected Value in One Year
Scenario 1:		100%	
No Regulatory Risk	$100	——————→	$115
Scenario 2:		50% ↗	$115 *(Good Outcome)*
New Regulatory Risk Assumed, but Commission Keeps Allowed Rate of Return Equal to Cost of Capital	$50	50% ↘	$0 *(Bad Outcome)*
Scenario 3:		50% ↗	$230 *(Good Outcome)*
New Regulatory Risk Assumed, but Commission Allows Regulatory Risk Premium of 115% over Cost of Capital	$100	50% ↘	$0 *(Bad Outcome)*

Cost of Capital = 15%

If investors have a 50 percent chance of a 100 percent loss, and if the risk is a coin toss (and therefore diversifiable), the stock price will fall to $50. The logic is that if investors face a coin toss with a $100 good outcome and a $0 bad outcome, the expected value is $50. (Note that

---

[23]   Note that this characterization makes the regulatory risk diversifiable in large portfolios. Thus, our conclusions in no way depend on the assumption that regulatory risk is a nondiversifiable risk that requires compensation via an increase in the cost of capital. However, as discussed below, that outcome is not precluded.

we are assuming no market imperfections or asymmetric information; investors know exactly what is going on.)

What happens to the DCF estimate? That depends on the particular practitioner of the theory, but one approach might be to assume that the expected dividend remains at $10, so the dividend yield is 20 percent.[24] If there is no loss, the *dividend* growth rate is still 5 percent, producing at best a 25 percent rate of return.[25] This seems like a high return, but is it enough?

The answer unfortunately is no. If investors are to voluntarily bear this risk, the issue is what rate of return has to be granted for the stock to trade at $100, the amount originally invested, after the risk is known but before the decision is announced. Investors have a 50 percent chance of losing $100 in book value and their 15 percent return, for a total downside loss of $115. Moreover, with the cost of capital at 15 percent, for the stock to trade at $100, the *expected* value in a year must also be $115. As shown in Scenario 3 of Figure 3-3, the outcome in the 'no

---

[24] Of course, in these circumstances, one would have to question the competence of an analyst who concluded that the $50 stock price embodied investor expectations that the company's most recent dividend would grow at 5 percent forever. Investors surely are more pessimistic than that, given the 50 percent risk of bankruptcy in one year. But a more realistic assessment of investor expectations would yield a better estimate of *the cost of capital*, not a better assessment of the allowed rate of return needed to give investors a fair opportunity to earn the cost of capital.

[25] Given the assumptions of the example, the true cost of capital remains at 15 percent because the risk of the environmental penalty is assumed to be fully diversifiable. Therefore, the DCF result of 25 percent overestimates the cost of capital, because investors no longer expect a dividend of $10 that will grow at 5 percent forever. In actual cases with regulatory risk, the dividend expectations will almost certainly no longer fit the assumptions that underlie the basic DCF formula, i.e., dividend yield plus growth) so a more complicated formula would have to be used to estimate the cost of capital. But if the right formula were found and the cost of capital estimated correctly, the result would be the analogue of the 15 percent in this example, not the analogue of the fair allowed rate of return.

loss' case must therefore be $230 per share,[26] which implies a fair allowed rate of return of *130 percent*.[27] Only if the new rate of growth were fortuitously estimated to be an unbelievable 110 percent per year forever could DCF get the 'right' answer with a 20 percent dividend yield (i.e., the answer that yielded the allowed rate of return that gave investors a fair opportunity to earn their cost of capital, not the answer that gave the 15 percent cost of capital that is the quantity the DCF model is designed to produce). Thus DCF estimates of the cost of equity in practice will *not* yield the fair allowed rate of return, even when they overestimate the cost of capital.[28]

---

[26]   The calculation is 50 percent x [Good Outcome Value] + 50 percent x [Bad Outcome Value] = [Rate Base] x (1 + [Cost of Capital]), or 50 percent x [Good Outcome Value] + 50 percent x [$0] = $100 x 1.15 = $115. The [Good Outcome Value] that fits this calculation has to be $230.

[27]   If we change the assumptions so investors get to keep the allowed earnings even if they lose the rate base, the fair allowed rate of return is 'only' 65 percent: 50 percent x [$165] + 50 percent x [$65] = $115 = [Rate Base] x (1 + [Cost of Capital]).

[28]   We are indebted to J. Stephen Gaske for pointing out that if instead the commission were to allow the 25 percent DCF estimate forever, it could turn out to provide an exactly compensating return or even to be too high. ("Use of the DCF Method for Determining the Allowed Rate of Return," working paper in progress.)

However, such a fortuitous result should not be interpreted as vindicating the DCF methodology as a correct measure of regulatory risk. The hypothesized risk is a one-time event, so the economically efficient solution is to compensate those who bear it with a high rate of return and then to let the allowed rate of return fall back to the cost of capital for future generations. This also is the only realistic approach, since regulators a decade or two from now can hardly be expected to grant a premium rate of return based on a DCF analysis today (and if they did, the problem of calculating the 'right' DCF number in a subsequent hearing would be even harder, because of the stock market's reaction to the policy). Reliance on regulators' using a 25 percent rate forever also produces a paradox, because future regulators are assumed to get the answer right, not because they believe that DCF gets the answer right, but because they explicitly account for the fact that past regulators used DCF to get an answer that was seriously wrong.

Having said all that, Gaske's analysis does point to the possible need for 'rate levelization' of the risk premium. A 130 percent rate of return in a single year would lead to an unacceptable spike in rates, so the logical alternative would be to

Nor is there any easy interpretation of the stock price to derive the fair target rate of return, because the stock price in the face of regulatory risk will reflect unobservable investor expectations of regulators' decisions. Thus, the $50 stock price in the above example could be used to infer the fair allowed rate of return only because we assumed we had gathered substantial information on perceived regulatory risk, beyond that normally used in a DCF application, from the investment community. Unless it is somehow possible to determine precisely what investor expectations about regulatory risk are, the observed price cannot be used to infer the fair target rate of return.

---

capitalize the extra 115 percent (over the 15 percent cost of capital) in the rate base and to amortize it over however long a period was needed. Note that this would approximate the competitive outcome in similar circumstances. For example, if 50 percent of oil wells are dry holes, successful wells will be worth twice their drilling and acquisition cost in a competitive equilibrium (tax and timing issues aside). More generally, see William B. Tye and A. Lawrence Kolbe (1992).

Figure 3-4
Allowed and Actual Rates of Return for Gas Pipelines

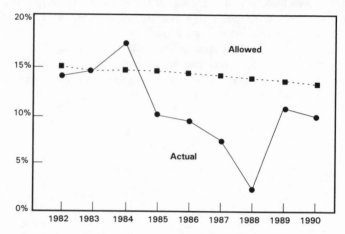

Sources:
Allowed ROE: INGAA Risk Survey
Actual ROE: Net Utility Operating Income Survey

In practice, there is no evidence that allowed rates of return have even approached the fair target levels. For example, Figure 3-4 shows that the rate of return allowed by the Federal Energy Regulatory Commission (FERC) for gas pipelines trended downward, even as gas pipelines were absorbing considerable 'take-or-pay' costs in the billions of dollars. The paradox is that these reductions in the allowed rates of return were occurring despite the fact that any casual observer would conclude the risk of extraordinary losses not only has existed, but that such losses have actually happened. Clearly such allowed rates of return did not reflect the extreme levels of asymmetric risk these carriers were facing.

Before leaving the DCF objection, it seems useful to address a related set of objections based on the 'comparable industries' test. Now somewhat out of favor, this method for estimating the cost of capital proceeds by finding an industry with comparable risk as the utility in question and taking its past realized accounting rates of return as the estimate of the cost of capital. Some critics suggest that this would provide a correct estimate of the necessary risk premium.[29]

This suggestion for calculating the necessary regulatory risk premium suffers the same problems as the DCF methodology. If the comparable industry used for the sample involved large numbers of investments over a long period of time, both successful and unsuccessful, it would, by definition, produce a downward-biased number for the necessary allowed rate of return for used and useful prudent investments.[30]

Changes in the Cost of Capital. Another loose end is whether the cost of capital itself might not go up in response to the regulatory risk. There is no reason why it should not in general. For example, modern finance teaches that risks that are hard to diversify, such as those responsive to general business conditions, increase the cost of capital (Brealey and Myers 1988, Chapters 7 and 8). It is easy to imagine that the risk of abandonment of a power plant under construction might be higher when times are hard than when the economy is booming. If so, the likelihood and perhaps the size of the expected loss due to plant abandonment could be higher during bad times. The cost of capital of a utility facing such risks may well be higher than that of one without such risks.[31]

If so, however, only the details of the above discussion change, not the conclusions. Consider the corporate bond analogy again. Suppose the company issuing the bond is in trouble because the economy is weak, so lenders expect less in the event of default, and the odds of default have

---

[29]　One reader privately communicated:

> The cost of capital is derived by observing market conditions in industries with similar risks. So if the risks in the utility industry are increasing, the way to find the appropriate cost of capital is to look at a slightly more risky industry.

[30]　The comparable earnings approach has serious defects as a cost of capital estimation method, too. See, for example, Kolbe, Read, and Hall (1984, Chapter 3).

[31]　The importance of nondiversifiable risk is the central intuition that underlies the Capital Asset Pricing Model.

increased. As a result, the company has had to promise a higher rate of return. Suppose as a result that the bond's expected rate of return (i.e., its cost of capital) also has increased. It clearly would be as unfair as it ever was to reduce the payoff to lenders because they expect less than they are promised. The fact that the cost of capital has gone up in no way offsets the requirement to pay lenders the full promised amount if the bond does not default.

Thus, regulators' recognition in the allowed rate of return of any cost of capital increase caused by the regulatory risk is *not* a sufficient remedy for regulatory risk. The fair allowed rate of return just has to be increased that much more, so investors have a fair opportunity to earn the new, higher cost of capital despite the regulatory risk.[32]

Having said that, however, it is important to reiterate the fundamental conclusions of the chapter. As shown in Figure 2-4, our findings in no way depend on the assumption that regulatory risk is nondiversifiable. Whether regulatory risk raises the cost of capital or not is an empirical issue to be decided on a case-by-case basis. But whatever the cost of capital is in a given case, the law of averages dictates the inexorable result of Figure 2-4: the target allowed rate of return on regulated investments without disallowances must exceed the cost of capital by an amount sufficient to offset the expected losses on regulated investments with disallowances.

Does the CAPM Logic Refute the Need
to Account for Regulatory Risk?

Another variant of the argument that the estimated cost of capital automatically provides sufficient compensation for investor risk reaches this conclusion via exactly opposite reasoning than the DCF objection.

---

[32]   The preceding discussion should reassure any readers who felt we were urging compensation for diversifiable risks as a general matter. It is only when such risks are *asymmetric*, like the default risk on a corporate bond, that they need to be taken into account in the target rate of return. And the only reason to do so is to permit investors a fair opportunity to earn the cost of capital itself.

The Capital Asset Pricing Model (CAPM) assumes that only nondiversifiable risks should be accounted for in the cost of capital.

According to the CAPM theory in its simplest form, 'beta' is a measure of the nondiversifiable risk of a stock. A beta of 1.0 is equal to the risk of a very large diversified portfolio of common stocks (i.e., the entire stock market). A beta of zero has no such risk, and might be represented by Treasury bills. The cost of capital is a linear function of beta.

Some critics have used these assumptions to conclude that the logic of Figure 2-4 is incompatible with the CAPM model. More specifically, their logic is that CAPM tells us to ignore asymmetric regulatory risks because they are fully diversifiable in a large portfolio of stocks. Among these diversifiable risks they would include the risk of regulatory disallowances of imprudently incurred costs, investments that turn out not to be 'used and useful,' and other past and future risks associated with uncertainty in the regulatory climate. These critics might argue that we are proposing to overcompensate investors by paying them a regulatory risk premium to bear diversifiable risks.

Thus, this logic for what we will call the CAPM variant goes as follows:

1. If high risks exist or have existed, they are of the regulatory variety;

2. Regulatory risks are diversifiable;

3. Diversifiable risks do not affect the cost of capital and hence need not be accounted for in the allowed rate of return; and

4. Regulators may therefore proceed as if these risks did not exist.

This is a different claim from that which says the DCF model yields a rate of return that, because the additional risk lowers stock prices, includes an additional risk premium which is fair compensation for these risks.

Clearly, both arguments cannot be right. Indeed, they raise the question as to whether one or both models are seriously deficient as estimates of the cost of capital. As long as one or the other is right, regulators need not be too concerned: both approaches claim regulators can accept the estimated cost of capital as the fair allowed rate of return. However, if *both* are wrong, regulators will need to reassess their current procedures.

Both, in fact, are wrong. Whether or not the risks are diversifiable is irrelevant to the need for a regulatory risk premium. The logic of Figure 2-4 does not necessarily reject CAPM as an appropriate methodology for measuring the cost of capital nor does it necessarily assume that regulatory and other asymmetric risks are not fully diversifiable. Indeed, the above coin toss example computed the necessary 130 percent rate of return on the assumption that the cost of capital was totally unaffected by the coin toss risk and was fully diversifiable at zero cost. Stated most simply, the logic of Figure 2-4 depends on the law of averages. It holds regardless of whether the CAPM is a complete explanation of how the cost of capital is determined.

If the regulatory risk were fully diversifiable and the CAPM were exactly right, the cost of capital would be unaffected by the introduction of new regulatory risk. But that would not imply that no premium over the cost of capital is required in the allowed rate of return, any more than a bond's default risk being fully diversifiable would mean bondholders would demand no default premium in the promised yield to maturity.

Indeed, the irrelevance of the diversification issue to the entire point can be proved by *reductio ad absurdum*.[33] Suppose that the CAPM model were absolutely infallible and in a particular situation showed that a utility's entire risk were diversifiable. In this case, its cost of capital would be equal to the risk-free rate. Suppose, in particular, that the regulatory risk were also entirely diversifiable. Would it make sense to give the utility the risk-free rate in the no disallowance outcome? Of course not, since the investor would still be faced with the prospect of a downside disaster. No rational investor would voluntarily commit

---

[33]   We are grateful to an anonymous referee for pointing out to us the properties of the limiting case of zero nondiversifiable risk as an illustration of the fallacy of the CAPM objection.

funds to such an investment (or diversified portfolio of such investments) even if the regulatory risk were 100 percent diversifiable.[34]

Does the Recognition of Regulatory Risk Depend on Imperfect Capital Markets?

The fourth objection to the conceptual framework seems to be a variant of the third. It is sometimes argued that one can ignore regulatory uncertainty by assuming that the "market operates efficiently." Our critics sometimes assert that the models for estimating the cost of capital rely on perfect markets. For example, critics assert that stock prices reflect all known risks, whether asymmetric or not. They then assert that these models are inconsistent with the need to account for regulatory and other asymmetric risks. Ergo, they conclude that the argument for a regulatory risk premium must rest on market imperfections to reach the conclusion.

Obviously, everything that has been said in support of the logic of Figure 2-4 depends in no way on hidden assumptions about market imperfections. The issue is whether models, such as DCF and CAPM, that are designed to capture capital market risk can be expected to capture asymmetric regulatory risk as well. The above counterexamples show that they cannot.

However, this objection seems to go deeper than mere theory. Critics claim that a regulatory risk premium is needed only if it is an unanticipated event not expected by investors, and thus already reflected in current security prices. But the general principles apply even when the risk is perfectly perceived by investors. Indeed, all of the examples in Chapter 2 assumed an agreed upon 'disallowance distribution' to show the need for a regulatory risk premium. This result is also readily seen

---

[34] Another referee has proposed an equally compelling example: suppose the disaster occurred with certainty. For example, suppose that *every* time utility investors put up $100, regulators disallowed $50 and put only $50 in the rate base. Under the CAPM, the cost of capital would be entirely unaffected. But if new capital is to be forthcoming, regulators would have to allow a rate of return high enough so investors could expect to earn the cost of capital *plus* the confiscated $50.

in our above coin toss example, where investors knew exactly what was going on[35] and yet the need for a regulatory risk premium over and above the cost of capital did not go away.

Some of our critics have gone to great lengths to demonstrate that investors have had adequate warning that rule changes would, in fact, happen. They attempt to demonstrate that regulatory disallowances were both theoretically possible and historically in evidence. Some observers note that an outright switch such as occurred in *Duquesne* is a rare event, and most disallowances were within existing rules.[36]

These respondents make a major distinction among: (1) disallowances and other asymmetric risks occurring within the existing rules, (2) disallowances occurring as a result of rule changes that were anticipated by investors, and (3) disallowances resulting from rule changes that were unanticipated by investors. These distinctions appear to be based on a

---

[35]   There is considerable literature on asymmetric information for regulated firms and regulators. See, for example, Wayne L. Lee and Anjan V. Thakor (1987). Our results emphatically do not depend on such asymmetric information, e.g., differences of opinion between investors and regulators as to the disallowance distribution.

[36]   *Duquesne* was the rare case of a 'smoking gun.' As one commentator put it:

> . . . it is exceedingly hard to determine whether many regulatory disallowances represent changes in the preexisting rules in a given state. In that sense, *Duquesne* was the unusual, easy case. The Legislature forced the agency to take an action that both characterized as a change in rules. In most cases, it is much more difficult to determine the preexisting "rule." There was virtually no "law" in most states concerning recovery of investments in canceled plants at the time utilities raised the capital to initiate the construction programs of the 70s and 80s.

> . . . given the murkiness of the preexisting law and its frequent dependence on discretion and findings of unverifiable facts, it is easy for a state to cause a utility enormous financial harm without appearing to change any preexisting rules.

belief that only the third class of risks should command a regulatory risk premium over the measured cost of capital.[37]

Thus, they conclude that, for the most part, such risks of disallowances or rule changes were already 'priced' in security markets and in measurements of the cost of capital. They interpret recent history of massive disallowances in the regulated industries as being a series of rare events accompanied by risk premia that were so small as to be 'lost in the noise' of the proper measure of the cost of capital.

This objection is thus a variant of the DCF objection addressed in detail above. We took the findings of fact in *Duquesne* to illustrate our principles, but they by no means require an explicit switch in rules such

---

[37]   An anonymous referee thus responded as follows:

> Any research into the effects of an adverse regulatory change must ask a fundamental question. Are investors realizing an unlucky event, but one for which they had long taken account? Or are investors being surprised by a change in the "regulatory contract" under which the securities were sold? The answer to this question is crucial. It is the difference, in Tye and Kolbe's analysis between a leftward shift of the bell-shaped curve, and merely landing on an unlucky portion of the curve. It is the difference between "taking property" and not, since investors would have accounted for the probability of the adverse policy change and would have demanded higher compensation accordingly.

> There is no clear *a priori* answer to this question. Some research suggests, however, that investors are good at assessing the hostility of the regulatory climate and adjusting returns (through the secondary market) accordingly. See S. P. Ferris, D. J. Johnson, and D. K. Shome (1986).

> Given the results of this research, it might be asked, "Have public utilities securities in general and in Pennsylvania in particular been adversely affected by the decision?" If not, investors may have long ago accounted for the probability of the imposition of the "used and useful" rule. If so, the change in rate base determination methodology took nothing from investors and the critique of the opinion rooted in constitutional law fails.

as the court found in *Duquesne*.[38] The basic theory applies regardless of whether a switch in rules occurred or not and regardless of whether investors anticipated correctly the event.

At the heart of all of these challenges, we believe, is an empirical objection: if regulatory risk is as big a problem as we say, how did regulated industries come to have the capital stocks they now possess? Does not the fact that people have invested for years under the specter of regulatory risk, indeed, that new equity capital was raised by electric utilities at the height of the period of disallowances in the 1980s, disprove our thesis? Maintaining our thesis in the face of this history leads some people to conclude we must be assuming either that investors are shortsighted or that some market imperfection exists.

In fact, our assumptions are quite different. To take the second point first, it is always possible to raise new capital for new investment if valuable claims on existing assets are thrown in. Buyers of new shares for troubled utilities in the 1980s did so at prices that gave them claims on the cash flows from old assets, which were supposedly the property of existing shareholders.

The issue of how the capital stock came to exist in the first place does require either shortsighted investors or a change of some sort. But that change need not come in the form of an explicit new regulatory rule, as in *Duquesne*. All that is needed is a change from a negligible probability of a material disallowance to a material possibility of a material

---

38   Buchmann (1991, 166) thus seems to be quarreling more with the Court's findings
      of fact rather than our analysis of the underlying economics when he states that:

           . . . the idea that *Duquesne* "changed" the law and vitiated a general
           regulatory "contract" is not supportable.

      We are in full agreement with his observation that an explicit change in the rules
      is not required to invoke the underlying economic principles in Chapter 2:

           While the Pennsylvania appeals involved a specific statue, enacted after
           commencement of the canceled units and subsequently modified on this
           issue, it is not at all clear that the Pennsylvania court might not have
           reached the same result anyway.

disallowance. Our experience suggests that such a change occurred in the 1980s. Nor do we believe we are alone in that view.

If the expected value of possible disallowances used to be small enough to be lost in the 'noise' of day-to-day business risks, investors could rationally have supplied the capital stock that existed at the start of the period of disallowances, even if it would not be rational to do so with the post-1980s level of regulatory risk.

In sum, we assume no capital market imperfections to reach our findings, nor do we believe these findings are inconsistent with the world as we see it. Indeed, the fear that electric utilities will be so reluctant to build new capacity that the United States will face a material shortage in coming years is reportedly a key finding of the National Energy Strategy (U.S. Department of Energy 1991, 86-87, 91, 93). Such a problem is exactly what our findings would predict.

Opportunities by Regulated Firms for Gains in Excess of the Cost of Capital

A number of commentators (such as Lyon 1991) have observed that stockholders in regulated firms do have opportunities for returns in excess of the cost of capital. Thus they state that the analogy of regulatory risk to junk bonds is overdrawn, because there is an absolute truncation of earnings for the bond, but not the stock. In this vein, one critic cites regulatory lag as a vehicle for returns in excess of the cost of capital and evidence at certain times of stock prices in excess of book values.

We agree with these observers that regulation does not flatly prohibit a rate of return to stockholders in excess of allowed returns. Indeed, a careful examination of the figures in Chapter 2 shows a distinct possibility of upside values. In this regard, we agree also that regulation does not impose an absolute truncation of shareholder returns as junk bonds on the rate of return to maturity.

However, a careful reading of our analysis will show that the conceptual framework is a statement about *expected values*. After the imposition of a new asymmetric risk, the expected value of the new distribution of values will be less than that of the old and less than the cost of capital if nothing else changes. The proposition does not say that possibilities on the upside are zero; rather, it says that the upside potential in the use of the standard model is inadequate in the presence of an asymmetric downside risk.

RESPONSES TO OBJECTIONS TO APPLICATIONS OF THE CONCEPTS

Does Recent Experience in Regulated Industries Represent a Retroactive Switch or Just Bad Luck?

The most fundamental objection to the applications of our theory is that what we have called traditional regulation already incorporated asymmetry. If so, we are told, recent massive disallowances incorporated no shift in the distribution; it was *always* something like that shown in Figure 2-4. After all, the reasoning goes, there was always some probability of a disallowance for imprudent investment inherent in the original *Hope* test of prudent original cost. How do we know that investors did not recently experience a rare event of a big loss that was explicitly recognized *a priori* by investors and accounted for in the cost of capital and allowed rates of return?

This is another facet of the objection that our results assume some market imperfection, but seems more to object to our finding that compensation above historical levels is required than to quarrel with our theory. Sometimes this argument seems motivated by concerns of fairness: investors were unlucky, not treated unfairly, so no changes need be made in the way regulation proceeds.

We then take this response as a plea that what we say may or may not be true in theory, but the problem of asymmetric regulatory risk was, or at least will be, *de minimis*. Since we have demonstrated in Chapter 2 that regulatory and other asymmetric risk is idiosyncratic to the particular circumstances, we must concede that our critics might well be correct in certain regulatory settings. But they in turn must agree that there will certainly be many cases, such as with *Duquesne*, where the Court found the facts support an undeniable conclusion that an expedient change in rules has indeed occurred (the State legislature passed a new retroactive law during the last week of the rate case). As Richard Pierce (1989) has noted, however, in many cases the prior rules for risk allocation were far from clear.

Given possible ambiguities, it is important to stress again that the above logic does *not* depend on an explicit change in the regulatory rules. The logic of Figure 2-4 is triggered under a *consistent* application of any of the possible ratemaking methodologies and does *not* require an explicit change in the rules. As long as there is any nonzero probability of an asymmetric downside loss *for any reason*, the upside outcome must exceed the cost of capital.

The source of the problem need not involve regulators at all. Take, for example, the case of the risk of 'stranded investment.' This consists of a once useful investment that has become obsolete through recent market or technological developments. Or perhaps unrealistic depreciation lives have prevented amortization of the investment over its useful life. Suppose that a large pipeline is built jointly by competing interests, each of which separately owns and prices its share of the facility. Demand for the facility is initially strong, but there is a nontrivial probability that gas reserves will decline, or demand for the throughput will abate, before the physical life of the facility has ended. It is anticipated that when this happens at some future point in time, a price war will break out and discounts for transportation services will drive rates down.

From the perspective of investment risk, such an event would be no different than if regulators had convened and declared part of the facility unused and useful and ordered a rate cut. The diagnosis of such asymmetric competitive risks is really no different than the diagnosis of asymmetric regulatory risks. Risks of other adverse events, such as the discovery of unexpected environmental cleanup costs, would also be analyzed via the same procedure (Hogan and Kolbe 1991, 34-37; Kolbe and Tye 1992, 18-20).

The point is that there is a large class of risks that create 'bad outcomes,' which affects the necessary allowed rate of return in the world of 'good outcomes' even if it does not affect the cost of capital (the expected value over all possible futures). All it takes to trigger the logic is that future states of the world be partitioned into 'Good Outcome' and 'Bad Outcome.'

We can expect regulated companies to respond to the level of regulatory risk they face. We believe our findings demonstrate that if this level is high enough, material changes are required in traditional regulatory rules.

What if Investors Buy After the Potential Regulatory Risk is Known?

Another comment by those who dispute our analysis raises the issue of switches in the degree of regulatory risk. Suppose, for example, that investors provide funds to a utility where the regulatory rules provide that all prudently invested funds may be recovered from ratepayers. Assume that the state legislature now passes a law prohibiting the recovery of investments in canceled plants.[39] The argument is that if someone buys the security after a new regulatory risk becomes known, that person will be fairly treated if the target rate of return just equals the cost of capital. Would it not give a windfall to current shareholders, a skeptic might ask, to reverse asymmetric regulatory policies once they are adopted?

Such a policy reversal might indeed provide a windfall for some shareholders, but we would argue that the policy reversal should be made anyway. One way to see this is to recognize that what might be a windfall for some shareholders would only restore others to their previous condition. The proverbial widows and orphans, long-time holders of utility stock, would have had every reason to be unhappily surprised as asymmetric regulatory risk grew in importance. If fairness to individual shareholders is to be the criterion, it would be unfair to long-time shareholders to *fail* to reverse seriously asymmetric policies.

---

[39] This example is of more than passing interest, because it corresponds to the Supreme Court's findings of fact in *Duquesne*.

The more fundamental response, however, is that the argument itself is just a way of saying that a stock is always fairly priced at the time someone buys it (absent insider trading or the like). But that does not mean regulators should select their policies to try to ratify that price. A stock could be fairly priced on Wednesday under a regulatory bargain without asymmetric risk, be fairly priced on Thursday after losing half its value in response to the announcement of a change in that bargain, and fairly priced again on Friday at the original Wednesday price after the commission in question announced it had reconsidered. The commission surely should not forgo that reconsideration because some investors happened to have bought the stock on Thursday. Courts have long recognized that regulatory policy has to determine the value of regulated firms, not the other way around.[40]

In the presence of asymmetric regulatory risk, a policy that equates the allowed rate of return with the cost of capital cannot be maintained in the long run, even if investors know about it in advance, because every new investment will be worth less than it cost.[41] Reconsideration of such a policy should not be held up by the fact that some investors bought shares before it was reconsidered.

---

[40]   For example, *Hope* states,

> "[F]air value" is the end product of the process of rate-making not the starting point . . . . The heart of the matter is that rates cannot be made to depend upon "fair value" when the value of the ongoing enterprise depends on earnings under whatever rates may be anticipated. 320 U.S. 601 [Footnote omitted.]

[41]   Note that it still may be possible to sell new stock to finance part of an investment under such rules, but the price new investors pay will be below the amount invested per share. The rest of the investment must come from the assets and earnings of the existing shareholders, diluting their claim on the new asset. But eventually the capital pool available to provide this cross-subsidy will be gone (or the managers who permitted it will).

Will Irrational Investors Save the Regulatory System?

Critics of the analysis of regulatory asymmetry often argue that the past rules may have been unfair, but investors were willing to supply capital anyway. The same will be true in the future, according to this point of view.

Reliance upon what we call the 'bigger fool' theory does not have a happy history in recent regulatory experience. Systematic, inadequate earnings in natural gas supply and railroads produced bankruptcies and shortages, and necessitated major regulatory reform legislation to correct poor incentives to invest. Management in the electric utility industry has often expressed disinterest in committing funds under the current regime (*Electric Utility Week* 1985, 3; Pierce 1991). Natural gas pipelines are shedding themselves of their gas supply contracts because of their recent massive losses on take-or-pay contracts (historically pipelines could only, at best, break even on recovering their gas supply costs).

The issue is not whether inadequate rates of return will produce zero investment. There will always be someone who may find a profit opportunity even in the most unfavorable investment climate. Rather, the issue is whether the amount of investment is optimal. Here the historical evidence is clear that inadequate profit incentives will ultimately produce suboptimal investment.

Regulatory Asymmetry and the Capture Model

Some observers cite various economic theories of regulation to conclude that the concept of regulatory risk is inconsistent with various 'capture theories' of regulation. According to the theory, regulators protect investors from hardship in exchange for a stable industry and various cross-subsidies to keep various constituencies happy (Posner 1974 and 1971; Stigler 1971; Peltzman 1976).

Pierce's (1989) survey of the history of recent regulation is instructive. He cites a jockeying for political advantage between investor and ratepayer interests in the utility field. The former group gained advantage during the 1950s and 1960s. As soon as downside disasters appeared, however, regulators swung over to favor ratepayer interests, where the situation remains.

In any event, recent scholarship[42] has provided data that question the capture thesis (Etzioni 1986). Recent literature alleges that the theory does not describe current practice, whatever its historical validity (Baron 1988).

However, the need for a regulatory risk premium over and above the cost of capital does not depend on one's position on the capture theory of regulation and does not require one to choose among these theories of regulation. Since the mechanics of the calculations do not require rule changes or regulatory expediency, one need not accept any particular theory of the origins of the regulatory rules or changes in such rules to accept the fundamental conceptual framework.

RESPONSES TO OBJECTIONS TO
OUR POLICY RECOMMENDATIONS

Have Changes in the Regulatory System or
Other Offsets Eliminated Regulatory Risk?

A number of commentators have claimed that recent regulatory changes have not been a one-way street. Regulatory reforms have given, in many circumstances, offsetting freedoms to compete and have introduced many economic efficiencies.

In a similar vein, other observers claim that regulated firms and investors have learned from past experiences. As a result, increased regulatory risk is claimed to be wholly retrospective.

These claims cannot be addressed except with a systematic study of the facts in the particular circumstances. Nevertheless, it is noteworthy that

---

[42]   We are indebted to an anonymous referee for providing insights into the connection between this literature and our central thesis.

all of the recent regulatory disasters have occurred in environments where the risks were said by many to be low prior to the discovery of the downside potential. Investors may well be skeptical of such claims in the future.

Difficulties in Implementation

A number of commentators have alluded to 'real world' problems in implementing a regulatory system that explicitly recognizes regulatory risk. We agree that the introduction of a large degree of regulatory risk compounds the already difficult problems of proper incentives for regulated firms. Alan Buchmann (1991) has added a number of practical difficulties to the list we originally compiled in Chapter 2.

However, it is important to recognize that such problems are the inevitable consequence of a regulatory climate embodying asymmetric risk, not a signal of fault in the economic analysis that diagnosed the problem. We analyzed the economic consequences of the regulatory system as we found it -- we did not invent that system. If there is a problem, the solution is to change the institutions to solve the problem, not to deny existence of the problem because facing it will make life difficult for regulators and regulated firms.

We readily accept the existence of many difficult problems in implementing these concepts. We hope that consensus on the conceptual framework and diagnosis can provide a basis for constructive suggestions for innovative solutions.

CONCLUSIONS

If asymmetric regulatory risks continue, some new approach is needed to compensate investors for those risks. One approach is to develop a new way to derive a target allowed rate of return on equity.[43] That is, a method might be developed to find the equity equivalent of the fair promised rate of return on a bond, not just the expected rate of return that the modern cost of capital estimation methods actually produce. The fair target rate of return will often be far higher than any ordinarily granted, and it will be very hard to calculate. For these reasons and for additional reasons discussed in Chapter 2, attempts to derive and implement fair regulatory risk premia will, at best, be extraordinarily difficult.

A far better solution to these problems is to eliminate the regulatory asymmetry entirely. One way to do so is to hold a symmetric regulatory bargain, in which ratepayers receive the benefits of unusually good outcomes and pay the costs of unusually bad outcomes. Another way, where possible, is to relax or eliminate regulation, so investors get to keep some or all of the fruits of unusually good outcomes, if they are to bear some or all of the costs of unusually bad outcomes.

What is at stake here is the appropriate level of incentives for future investment in regulated utilities. This chapter has not provided a method for deciding what the appropriate level of investment should be. But it is clear that the expected rate of return is not equal to the allowed rate of return under asymmetric regulation. Under these conditions regulators will not be providing the incentives they think they are giving if they continue to practice traditional regulation and ignore the effects of asymmetry.[44]

Consequently, regulation as sometimes practiced in the 1980s must change. Capital already on the table can be confiscated by adoption of unfair rules, but there is no way to force investors to put new capital in

---

[43]   If the allowed rate of return cannot be adjusted, perhaps an explicit insurance premium could be adopted for some classes of risks. From an economic perspective, the Gas Inventory Charges (GICs) now proposed by the FERC for gas pipelines are an example. However, it is likely to prove very difficult to calculate an economically correct GIC.

[44]   The result will be in effect a 'reverse Averch-Johnson effect,' or an incentive to underinvest (Train 1991, 104-105; Pierce 1992).

the game afterwards. Regulators have two feasible options: convincingly forswear asymmetric penalties in the future or grant before-the-fact compensation for assuming the risk. If neither option is selected, either new institutions will have to evolve to provide regulated services, or those services will not be provided well.

Chapter 4
Sources of Asymmetric Risk
in Regulated Industries

Incentives to discount are strong where traditional cost-of-service rates provide substantial differences between pipelines. For example, the two interstate pipelines serving Southern California from West Texas and New Mexico -- El Paso and Transwestern -- have significantly different cost-of-service tariffs since El Paso is an older, more fully depreciated pipeline while Transwestern is a newer line (Jensen Associates 1990, 49).

Some background in the relevant economic principles is necessary for the analysis of the following Chapters 5 through 9 on gas pipelines to make sense. In particular, this chapter:

1.   Contrasts the prices charged for capital investments under competition and under traditional regulation; and

2.   Explores the kinds of risks that do and do not affect the cost of capital.[1]

CAPITAL CHARGES UNDER REGULATION
AND COMPETITION

Prices behave very differently over time under traditional rate regulation and under competition. This can create serious difficulties for firms that face both regulatory and competitive constraints. Essentially, such firms can end up being worth 'the lower of cost or market'; competition can restrain their earnings when regulation would permit higher earnings, and regulation can constrain their earnings when competition would permit higher earnings. The result can be inefficient purchases by customers and inefficient -- or inadequate -- investment by regulated companies.

---

[1]   As defined in Chapter 2, the 'cost of capital' for an investment is defined as 'the expected rate of return in capital markets on alternative investments of equivalent risk.'

The reasons for this difference turn out to be linked to the choice of a rate base methodology and a depreciation schedule under regulation. This section first describes the generic problem, and then discusses some of its specific implications for the gas pipelines and other industries.

The Generic Problem

Economists usually view rate regulation as an attempt to approximate the results of competition. Two characteristics of competitive markets in equilibrium govern the capital charges that are implicit in competitive prices:

1.　　Investors expect a fair opportunity to earn a rate of return on investments that is just equal to the cost of the capital they supply; and

2.　　The price customers pay *at any point in time* is independent of the age of the assets owned by the supplier from whom they buy.

The first of these is a standard goal of rate regulation, although as discussed in the last part of this chapter, achievement of this goal is not as straightforward as it might seem. The second characteristic, however, is usually not even mentioned in regulatory hearings.[2] Yet a moment's thought will reveal how commonplace the condition is in competitive industries. For example, the price of tomatoes does not depend on whether the farm was bought in *1*891 or *1*991. If this condition did not hold true, new entrants and established firms would charge different prices for the same product at the same location at the same point in time. There would be a bin at the supermarket with cheap tomatoes from old farms right next to a bin with identical, but far more expensive,

---

[2]　　An exception is regulation of oil pipelines, where the FERC has adopted a rate base that is trended for inflation, in part because of the intense competition that exists in the oil pipeline industry. See Opinion No. 154-B, *Williams Pipe Line Co.*, 31 FERC (CCH) 61,337, *modified,* Opinion No. 154-C, 33 FERC (CCH) 61,327 (1985).

tomatoes from new farms. Obviously, this does not happen in a competitive market.

Conventional original cost ratemaking with straight-line depreciation, however, implies rates that vary considerably with the vintage of the assets employed. Identical pipelines built 20 years apart would have very different capital charges,[3] and customers would pay considerably less for transportation over the older pipeline. This simply would not occur under competition.

If regulation charges too little for the services of old assets, but investors still expect to earn the cost of capital over the life of the asset, then logically, regulation must charge too much for new assets. This turns out to be the case. Regulation results in a significant 'front-end load' relative to competitive prices. Figure 4-1, which ignores taxes for simplicity, contrasts the time paths of capital charges on a regulated and competitive asset, if the cost of capital is constant over the entire period.

The total capital charges in the figure consist of depreciation, the return *of* capital, and earnings, the return *on* capital. As depicted in Figure 4-2, ratemaking depreciation is typically straight-line, which implies an equal depreciation charge in every year. Regulated earnings therefore decline linearly, as straight-line depreciation erodes the rate base. Total capital charges on a regulated asset are much higher early in the asset's life than later on.

The other factor that must be considered in competitive prices is inflation, which obviously has a profound effect on the economically appropriate time path of rates. Use of a book value, historical cost rate base materially exacerbates the front-end load due to straight-line depreciation: since regulated investors cannot keep inflation-driven appreciation in the value of the rate base, they must instead receive inflation compensation through a higher allowed rate of return. This happens when regulated firms are allowed a *nominal* rate of return on a *historical* cost asset base, instead of a *real* rate of return on a *current* cost asset base, as happens

---

[3]   As used here, 'capital charges' consist of the sum of depreciation, operating earnings, and taxes.

Figure 4-1
Capital Charges (Earnings and Depreciation) Under Alternative
Capital Recovery Time Profiles

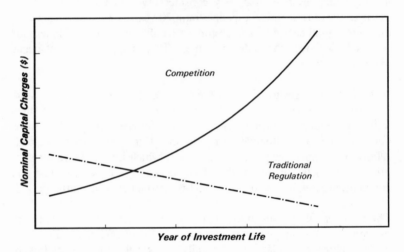

in competitive markets. Recent bouts of inflation have become a major motivation to search out more economically appropriate treatments of capital costs.

The pattern under competition is very different from that under regulation. Since the competitive market price at any point in time is independent of the vintage of assets owned by any particular competitor, the natural starting point is a level time path in *real* terms, i.e., after the effects of inflation are taken out. (The average price is level in 'real terms by definition -- that's how a price index is constructed; all else equal, capital charges will be level in real terms, too.[4]) Of course, if

---

4   Of course, all else may not be equal. The competitive time path of capital charges may not be level in real terms. Technical change may imply that real prices for a particular industry are declining over time. Alternatively, real increases in construction costs may imply real capital charges that rise over time. And even if real *prices* are constant, variations in output over time may imply that real *capital charges* vary. These factors and others can argue for more rapid recovery of capital

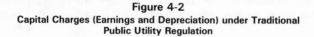

Figure 4-2
Capital Charges (Earnings and Depreciation) under Traditional
Public Utility Regulation

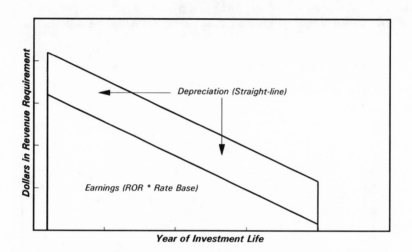

Year of Investment Life

capital charges are level in real terms, they rise at the rate of inflation in nominal terms. This is responsible for the upward slope to the 'competition' line in Figure 4-1.

A level real path requires total capital charges, but *not* the individual components of capital charges (i.e., earnings, depreciation, and taxes) to be constant in real terms. The principle of level real rates (again ignoring taxes) is illustrated in Figure 4-3, which, unlike the previous two graphs, is in dollars of *constant* purchasing power. The figure shows that the two components of capital-related expense are structured to

---

than under level real capital charges. See William B. Tye and A. Lawrence Kolbe (1992) for discussion of these points.

produce the same total amount each year in constant dollars.[5] The 'deceleration' of depreciation (relative to straight-line) exactly offsets the declining return on capital to produce a level annual cash flow over time.

Figure 4-3
Real Capital Charges (Earnings and Depreciation) with No Decline in Demand or Technical Change

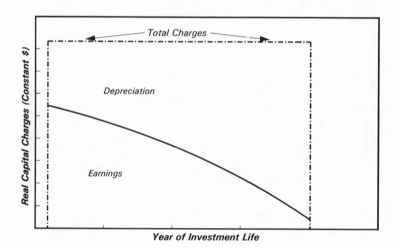

The problem with straight-line depreciation is seen most clearly by comparing the mortgage-like payments in Figure 4-3 with what would happen if homeowners had to contend with conventional utility ratemaking in Figure 4-2. Even though straight-line depreciation is level, total capital-related charges are heavily front-end loaded. To achieve level real capi-

---

[5]   If this graph were instead for payments in nominal dollars, it would look like the payment stream on a fixed-rate home mortgage. The return of capital (depreciation) is like the principal repayment component, while the return on capital (earnings) is like the interest component. As any homeowner with a traditional mortgage recognizes, the amortization of the principal (depreciation) is relatively modest at first and increases over time.

tal charges, the depreciation schedule must be decelerated considerably from the one usually used under regulation.[6]

The long lives of many capital-intensive investments such as pipelines amplify the distortions caused by prices based on original cost rate bases. In industries with rapid asset replacement, the book value of assets will stay closer to their true economic value, simply because less time passes before the assets are retired and new assets purchased.

The discussion to this point has considered single assets. But actual regulated companies are mixtures of assets of many different vintages. It is natural to ask whether the front-end-load/back-end-shortfall problem is eliminated in that case.[7]

The answer is that one or the other condition may dominate the company's rates, but that those rates will (virtually) always differ from the rates that would exist under competition.[8] In general, the issue is whether 'new' or 'old' assets dominate the rate base. Even if assets are added continuously at a constant growth rate, the most favorable assumption for the case that aggregation eliminates the problem, competitive and regulated rates will differ unless that growth rate equals the real cost of capital and the asset-specific rate of inflation in this industry is equal to the

---

[6]   Figure 4-3 also demonstrates the point that a desired time profile of total capital charges does not imply an identical time profile of every particular cost component. Figure 4-3 illustrates that economic depreciation is a residual, or 'fudge factor,' necessary to produce level real rates or whatever trend of real rates is desired. Indeed, in nominal dollars in times of high inflation, early economic depreciation on a new asset may even be *negative*. In general, either depreciation or the time profile of rates can be specified in advance, but not both. Whichever one is specified implies a unique value for the other. Thus, level real rates imply a (real) depreciation schedule like the principal repayment schedule on a home mortgage, while a straight-line depreciation schedule implies a time path of capital charges that declines linearly.

[7]   In the gas pipeline industry the result is called 'rolled-in' tariffs. Existing customers may object to new investment to serve new customers and demand instead 'incremental tariffs.'

[8]   Support for the points made in this section about asset aggregation and the effect of differing growth rates may be found in Stewart C. Myers, A. Lawrence Kolbe and William B. Tye (1985).

general rate of inflation. If the growth rate of such companies exceeds the real cost of capital, the companies will tend to have a permanent front-end load, and if it falls short of the real cost of capital, a permanent back-end shortfall.

Similarly, steady-growth companies with asset-specific rates of inflation that exceed general inflation will tend to have a permanent front-end load, unless the growth rate is very low. Steady-growth companies with asset-specific rates of inflation that fall short of the general rate of inflation will tend to have a permanent back-end shortfall, unless the growth rate is very high.

In reality, of course, companies do not have steady growth rates and asset-specific rates of inflation are sometimes higher and sometimes lower than general inflation. Thus, the differences between regulated and competitive capital charges definitely exist in the real world.

Implications for Gas Pipelines

When inflation is rapid, original cost regulation leads to prices to consumers that are far out of line with what a competitive firm would charge. Some gas pipeline rate bases, for example, are close to fully written off, yet remain 'used and useful.' The Sea Robin Pipeline is an example. Sea Robin's 1989 gross plant was $253,144,935, while its net plant was $4,888,194 (U.S. Department of Energy 1989). Competitive firms would obviously charge much more for the use of such assets than traditional regulation would permit. The underpricing of these companies' assets will distort decisions by producers, pipelines, and customers, who cannot be expected to make economically efficient choices if faced with the wrong prices. In the extreme, the services of a fully depreciated gas pipeline still in use would be virtually free, despite the obvious fact that the asset has economic value that has to carry a price tag if sensible decisions are to be made.

Such problems exist even for companies that are regulated monopolies. Still more serious problems arise when the pipeline is constrained by both regulation and competition. In this case, a new pipeline may find its rates constrained by competition early in its life, while regulation constrains its rates late in its life. The situation is depicted in Figure 4-4, which adds a heavy line to Figure 4-1 to show the effects of the two simultaneous constraints. The end result can be assets worth 'the lower of cost or market,' to borrow the accounting phrase, a serious problem for investors.

Figure 4-4
Capital Recovery May Be Impossible for Firms Facing both Regulation and Competition

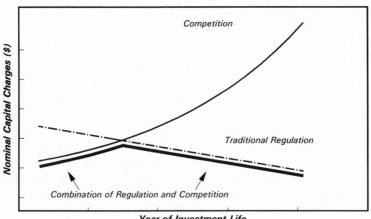

Investment may not proceed under such circumstances unless the pipeline is able to roll the new investment in with an old rate base and charge customers an average amount that satisfies regulators and still is at or below the prices competitors charge. New investments will be inadequate or misplaced where such averaging proves infeasible, and incentives to maintain and improve existing investments will be distorted.

Customers as a group lose under such mixed systems, too, not only because useful pipeline investments may not be made, but also because it may pay some customers to route gas in inefficient ways. For example, it may pay to choose a roundabout route on an old system over a direct route on a newer system. Such inefficiencies distort rates on both systems and waste resources.

Thus, there can be substantial differences between regulated and competitive prices. Regulation permits higher prices than competition early in an investment's life, while competition permits higher prices later in the asset's life. (Recall Figure 4-1.) A company that faces both constraints can find its earnings below the approved regulatory level early on, because of competition, while regulation keeps earnings low afterward, despite the fact that competition would now permit higher prices. A company that faced such constraints for a long period cannot survive, because investors eventually will refuse to supply the necessary capital to keep it going.

OTHER RATE-REGULATED INDUSTRIES

This book is a case study of regulatory change and regulatory risk in a rate-regulated industry. It focuses on the gas pipeline industry, which easily has seen as much change in the rules that govern its operating conditions and financial health as any industry that remains subject to rate regulation. The experience of the gas pipeline industry illustrates general economic principles, summarized above, which rate regulation sometimes overlooks to the detriment of current investors and future customers.

However, the difference between regulated and competitive capital charges and the effect of asymmetric payoff structures are important for other rate-regulated industries as well. To put this book in context, this section provides a brief discussion of how these principles have affected other industries.

Electric Utilities

Except possibly for gas pipelines, the electric utility industry is the leading exemplar of the effects of the economic principles that are the focus of this book. The electric industry has experienced fewer explicit major rule changes than gas pipelines, perhaps in part because it is regulated primarily by state commissions rather than a single federal commission. However, it has been as affected by the above principles at least as much as gas pipelines.

The electric industry for many years benefited from scale economies that made new, larger power plants cheaper than old, smaller ones. Under these conditions, the early capital charges on a new plant under regulation are close to the late capital charges on an old plant. Moreover, rapid growth in the demand for power meant new plants came on line frequently, which meant rates usually reflected the average of plants of many different vintages. The result was electric rates that fell over time, without sharp changes. Customers were happy, and investors were as well (Joskow 1974).

This changed in the 1970s, when energy price shocks sharply slowed the growth in demand just as the major engineering scale economies in plant construction were exhausted. Relatively suddenly, the unit cost of new plants became more expensive than for old ones, and new plants needed to come on line far less frequently. The result in the late 1970s and much of the 1980s was 'rate shock': a dramatic contrast between the capital charges for new electric plants and old ones, in circumstances where the new plant represented a substantial fraction of the dollar value of the existing rate base.

Customers naturally objected to such sharp price increases, since competitive prices do not display such sharp movements even when costs are increasing rapidly. (The problem under regulation is not only that costs are increasing, but also that old assets are underpriced and new assets overpriced, relative to the prices that would be charged for the same assets in competitive markets.) Regulatory commissions and state governments reacted to the requested rate increases in several ways. One result was the 'phase-in,' in which rates on the new plant were not set in the traditional manner, but rather were deferred over a period of years. This has the effect of making early regulated capital charges more like competitive ones.

Commissions also investigated the prudence of management decisions for plants that cost much more than originally anticipated. Such reviews often led to a disallowance of part of the construction cost from the rate base.[9] 'Used and useful' statutes, which said that the cost of plants not used to supply electricity could not be recovered, served a similar function for plants that were abandoned before completion due to slow growth in demand and rising costs. The resulting losses greatly exceeded anything in the previous history of the industry, and the effect has been to make electric utility managers reluctant to commit funds in advance of proven demand.[10]

The massive construction projects of the 1980s are now past, and the industry recently has been relatively quiet and profitable. However, the principles that are the focus of this book are likely to become important again in the relatively near future. Demand, while slowed by the current recession, has generally been growing faster than industry forecasts. Moreover, in recent years, competition in the electric utility industry has increased sharply, and legislative proposals to increase it still more seem likely to pass. Unless the issues of regulated versus competitive capital charges and regulatory asymmetry are faced and accommodated, it is likely that the electric industry will enter another difficult period in the mid to late 1990s.

Oil Pipelines

Oil pipelines are noteworthy because they represent the one industry where the commission with jurisdiction (the Federal Energy Regulatory Commission) has explicitly attempted to deal with the problem of regulated versus competitive capital charges. Oil pipelines have long been a highly competitive industry. Almost all oil pipelines are built as part of an integrated effort to get crude oil to refineries or petroleum products to market. Companies are free to build oil pipelines whenever they can get the requisite rights of way and environmental permits, without prior permission of the FERC or its predecessor for oil pipeline regulation, the Interstate Commerce Commission. The level of risk that pipeline owners

---

[9]   According to Standard & Poor's (1991, 16), prudence-based disallowance totaled over $10 billion in the 1980s.

[10]   Standard & Poor's (1991) conclusion in this regard is consistent with our own experience with utility managers.

have been willing to bear has reflected this integrated approach, and there have been oil pipeline projects that stayed empty for several years after completion or that simply failed entirely.

If this type of market is to persist, the distortions between regulatory and competitive capital charges must somehow be overcome. The ICC for many years regulated the industry with a loose hand, using a 'fair value' rate base that, like competitive assets and unlike traditional Original Cost rate bases, grew in value with inflation. When jurisdiction passed to the FERC upon its creation, an ongoing proceeding over the merits of the ICC's regulatory approach passed on also. The issue of what rate base to use for the industry was at the heart of this proceeding.

After a substantial delay that included the remanding of an earlier decision to let the ICC's fair value method stand, the FERC in 1985 established a 'Trended Original Cost' rate base for the equity portion of oil pipeline rate bases in Opinions 154-B and 154-C.[11] Under this approach, the original cost of the equity share of the pipeline assets is trended for inflation. In exchange, the rate of return allowed on the equity portion of the rate base is reduced by the inflation premium in the cost of capital. The result is a time-pattern of capital charges that more nearly replicates that implicit in competitive prices.

The problems that Original Cost regulation creates for industries subject to substantial competition were a chief basis for the FERC's decision to use a trended rate base. The fact that Original Cost is still used for the debt portion of the rate base, however, means that differences between regulated and competitive capital charges will still arise. Thus Opinions 154-B and 154-C are only a partial response to the differences between regulated and competitive capital charges. Moreover, the fact that the FERC is regulating oil pipelines with a tighter hand than the ICC raises the problem of regulatory asymmetry in a way that traditionally has been absent from this industry. The result is likely to be to discourage new

---

[11] Readers should be aware that the authors of this book were involved with this case. The Trended Original Cost approach was based on the Verified Statement of Stewart C. Myers in *Williams Pipe Line Company*, Docket No. OR79-1, *et al.*, 1979. However, the Myers testimony recommended trending all of the rate base, not just the equity share.

oil pipeline investment relative to the amount that would occur under a symmetric system with a fully trended rate base.

Telecommunications

The principles emphasized in this book might not seem to be as important for telecommunications. In general, the regulated portion of the telecommunications business has benefited from rapid technological change, which has continually reduced the cost of service. Telecommunications is reminiscent of the electric industry of the 1960s, where continually falling costs can mean the discrepancy between regulated and competitive capital charges is not as important. This reduces the scope for asymmetric regulatory policies as well.

However, even telecommunications companies have not been immune. The problems for telecommunications arise when competition erodes a company's ability to continue to charge approved regulated rates. The inability to continue to recover a high proportion of cost from the long-distance business, which ultimately led to the AT&T breakup and the current competitive market for long distance, is certainly a leading example. Perhaps the best example of asymmetry resulting from the interaction of regulation and competition is AT&T's write-off, at the time of the breakup, of billions of dollars worth of technologically obsolete equipment that regulators had not permitted the company to depreciate sufficiently rapidly.

Another exception is long-distance service in high-cost areas such as Alaska, where the need to keep affordable service available for remote, very high-cost, and often economically depressed areas has created opportunities for competitive entry in more populous areas.[12] Deregulation of the long-distance business in Alaska, unless implemented with care, will expose the initial carrier to an asymmetric risk of loss.

---

[12]   Readers should be aware that one of the authors (Kolbe) has done work for the original long-distance provider in Alaska -- Alascom, Inc.

Thus, even with cost-reducing technological change, the combination of regulation and competition can leave rate-regulated firms vulnerable to the concerns that are the focus of this book.

RISKS THAT DO AND DO NOT AFFECT
THE COST OF CAPITAL

Managers naturally focus on the risks in the business environment, on 'things that can go wrong.' Financial economists focus instead on capital markets, on the risks that affect the rates of return investors require. Some risks of intense concern to managers are of far less concern to investors, or at least of far less concern to the large, active, well- diversified investors whose trades often set security prices.

This section considers the kinds of risk that increase the rate of return investors require to commit their funds voluntarily. To address this question, we must distinguish chiefly between *diversifiable* and *nondiversifiable* risks. A more complete discussion of these and related points is in Appendix A.

The starting point for the rate of return investors require is the *cost of capital*, defined as *the expected rate of return in capital markets on alternative investments of equivalent risk*. Thus, the level of risk is intimately related to investors' required rate of return. But not all risks matter equally to investors. The basic point may be illustrated with a perhaps whimsical example.

Suppose you go to a gambling casino and decide to 'invest' $1,000 playing roulette. The range of possible rates of return on your $1,000 is high. There is a good chance you will end up with nothing of your $1,000 left, for a return of -100 percent. On the other hand, you might have an incredible run of luck and end up winning $10 million, for a return of +1,000,000 percent. Your *expected* rate of return is negative, because the odds on roulette favor the house, but you may decide the slim chance of enormous upside potential is worth the considerable (100 percent) downside risk. In any event, the volatility of your possible rates of return on an evening's gambling, viewed before you begin, is very high.

What is the casino's risk from roulette? For your time at the table, it is the mirror image of yours. But you are unlikely to be the only player that night at that table. And even if you were, the casino has other tables going at the same time. And those tables run not only the night you are there, but night after night, year after year. The casino's risk from roulette is thus very small.

It is small because the returns from successive spins of a roulette wheel and simultaneous spins of different roulette wheels are *uncorrelated* with one another. The more spins, the more predictable the house returns are for a given amount of money bet. The casino is *diversified* against risk from roulette, because good luck on one spin is offset by bad luck on another. For a given sum bet, the house can predict its rate of return from roulette over a year down to a very narrow range.

Does this mean gambling casinos are a risk-free investment? The widely publicized financial problems for some Atlantic City casinos suggests they are anything but risk-free. The casino's risk comes not from roulette, but from the attitudes of gamblers. Do they feel rich and optimistic, or poor and pessimistic? Do they expect to do well next year in their jobs and investments or not? Do they make a trip to Atlantic City or stay home and play the state lottery? If they come, do they bet a lot or a little? Casinos cannot diversify against a recession that impoverishes gamblers before they come or makes them want to stretch their remaining money as far as it will go.

This distinction between diversifiable and nondiversifiable risk is crucial in identification of what risks matter to investors. Diversification consists of the combining of assets into portfolios. The casino's roomful of roulette wheels playing night after night diversifies the risk of individual spins of a wheel, and the holding of a portfolio of stocks diversifies much of the risk of individual stocks. Random good news for one stock is offset by random bad news for another, stabilizing the value of the portfolio. By the time ten to fifteen randomly selected stocks are added, a material portion of the diversifiable risk of the individual stocks is eliminated.

The impact of a portfolio on diversifiable risk is illustrated in Figure 4-5, which plots the 'standard deviation' of a portfolio as more stocks are

added.[13] With one stock, the portfolio has a level of volatility equal to
the stock's. The portfolio's volatility declines as stocks are added, rapidly at first and then more gradually. Eventually the portfolio's volatility
'bottoms out.'

Figure 4-5
Impact of Number of Stocks on a Portfolio's Risk

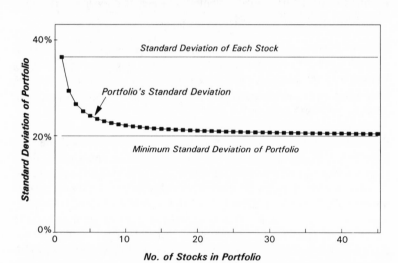

No. of Stocks in Portfolio

Where it bottoms out depends on the degree of correlation among the
stocks in the portfolio. If the stocks' returns were completely uncorrelated with one another, like spins of roulette wheels, the bottom would
be a level of volatility of (virtually) zero, implying no uncertainty at all
about the rate of return. (For example, the proportion of heads in a million, or a billion, flips of an honest coin will (virtually) always be 50
percent to a large number of decimal places.) Thus, if risk is entirely
diversifiable, it (virtually) does not exist for well-diversified investors.

---

[13]   Standard deviation is just a statistician's way to measure the amount of volatility.
The lower the standard deviation, the less volatile the portfolio's returns.

The cost of capital therefore will *not* include a premium for diversifiable risk.

For stocks, however, the volatility of a portfolio bottoms out at a positive level, indicating the portfolio has risk no matter how many stocks are added. That this must be true can be seen by even a casual reading of the financial pages. The stock market as a whole does move up and down, sometimes dramatically, which indicates that even a portfolio of *all* stocks would still be risky.

The risk that cannot be eliminated even in large portfolios is known as *nondiversifiable* or *systematic* risk. This is the risk that exists even for well-diversified investors, and therefore the cost of capital will include compensation for nondiversifiable risk if investors are risk-averse (as the evidence indicates they are).

People sometimes object that total risk, not just nondiversifiable risk, must matter to undiversified investors. There is no doubt that it does; if an investor owns only one stock and it does poorly, the investor cares. But the relevant question is whether a stock with high total risk but low nondiversifiable risk would be priced lower than a stock with the same nondiversifiable risk but lower total risk, to offer a premium rate of return on the first stock for the high diversifiable risk. In other words, can an undiversified investor expect to buy stock at a price that offers compensation for diversifiable risk?

The answer has to be no, because there is only one price for a stock. If a company's shares were priced especially low because it had especially high diversifiable risk, it would appear to be a bargain to well-diversified investors. Those investors could add the stock to their portfolios and never notice the diversifiable risk. There are many large, active, well-diversified investors in the market. They would snap up the bargain-priced (to them) stock, thereby driving up its price. In the end, the stock would be priced so it offered fair compensation for its risk to these investors (its nondiversifiable risk) and no compensation for diversifiable risk.

Thus, when the issue is the extent to which a risk affects a pipeline's cost of capital, the answer needs first to address whether or not the pipeline's investors can diversify that risk.

ASYMMETRIC RISK AND THE ALLOWED RATE OF RETURN

The message is that some risks that pipeline executives worry about do not affect the pipeline's cost of capital. However, that fact does *not* necessarily mean that regulators can ignore these risks when setting the allowed rate of return, as shown in Chapters 2 and 3. The logic of these chapters can be completed by now showing that it applies to any situation of asymmetric returns, not just regulatory risk.

It is now widely accepted that regulators should seek to establish rates such that investors can *expect* to earn their cost of capital on their investments (or at least on their prudently incurred, 'used and useful' investments). The reason is simple: if the *expected* rate of return on investment in the regulated industry is equal to the *expected* rate of return on alternative investment of equivalent risk, i.e., the cost of capital, the present value of the future cash flows investors *expect* will equal the amount they invest.[14]

In practice, regulators try to implement the principle that investors should expect to earn the cost of capital by equating the allowed rate of return with their best estimate of the cost of capital. However, it turns out that the mere equating of the allowed rate of return to the cost of capital need not, and in recent years *has* not, permitted regulated investors to expect to earn their cost of capital. Moreover, this problem is due not to any defect in the way the cost of capital is estimated nor to a systematic bias in estimation of the other costs of service. *Rather the problem arises from an innate flaw in the prescription to equate the allowed rate of return to the cost of capital.*

---

[14] In this context, 'expected' is used in the statistical sense, i.e., the value such that the probability-weighted average of the possible values above this level is exactly balanced by the probability-weighted average of the possible values below this level.

In short, there is a need for a fundamental reform in the way regulators think about compensating investors for the risks they bear. Therefore, it is essential that the reader understand and share the logic of the analysis that follows. The analysis demonstrates a vital distinction between the *allowed rate of return* and the *cost of capital*, which are frequently treated as synonymous. The allowed rate of return may have to differ from the cost of capital because of *asymmetry* in the distribution of returns investors face.

Asymmetry may be illustrated most easily by the example of the risk of material regulatory cost disallowances. Over the last decade, there has been a growing tendency in regulatory decisions to force utility shareholders to bear the burden of bad outcomes for decisions made under uncertainty. Examples include the partial or total exclusion of a power plant from an electric utility's rate base, and the requirement that shareholders bear some or all of the cost of buying out a gas contract that failed to foresee the downturn in world energy prices.

The source of the problem need not involve regulators at all, however. Take, for example, the case of the risk of 'stranded investment.' This consists of a once useful investment that has become obsolete through recent market or technological developments. Or perhaps unrealistic depreciation lives have prevented amortization of the investment over its useful life. Suppose, for example, that a large pipeline is built jointly by competing interests, each of which separately owns and prices its share of the facility. Demand for the facility is initially strong, but there is a nontrivial probability that gas reserves will decline or demand for the throughput will abate before the physical life of the facility has ended. When this happens it is anticipated that a price war will break out and discounts for transportation services will drive rates down.

From the perspective of investment risk, such an event would be no different from the case where regulators convened and declared part of the facility unused and useful and ordered a rate cut. The diagnosis of such asymmetric competitive risks is really no different from the diagnosis of asymmetric regulatory risks. Risks of other adverse events, such as the discovery of unexpected environmental cleanup costs, would also be analyzed via the same procedure (Hogan and Kolbe 1991).

The point is that there is a large class of risks that create 'bad outcomes,' which affects the necessary allowed rate of return in the world of 'good outcomes' (the upside branch in the figures we present below) even if it does not affect the cost of capital (the expected value over all possible futures). All it takes to trigger the logic is that future states of the world be partitioned into 'Good Outcome' and 'Bad Outcome.'

The basic character of the risks that affect the necessary allowed rate of return but not necessarily the cost of capital is *negative asymmetry*. Regulators will not see the full effects of these risks when the cost of capital is measured and so may set allowed rates of return that do not compensate investors for the risks they bear. In effect, some risks are like insurance losses, large and without an offsetting opportunity for gain.

Regulators must make some accommodation to these risks, if capital is to be attracted to the industry in the long run. For example, the allowed rate of return can be set above the cost of capital. Alternatively, the allowed rate of return could be left equal to the cost of capital and an explicit cost of service item, a form of 'insurance premium,' could be added as compensation for the type of risk described below. For example, this is one way to think of the function of Gas Inventory Charges, at least in principle.[15] Effectively, such a cost item would be a fee for providing certain asymmetrically risky services, just as an insurance carrier is compensated for casualty risk.

This is a standard feature of competitive markets. Competitive firms that make investments with a high probability of failure expect to earn enough on the successes to pay for the failures. There are many examples. Pharmaceutical companies do research on many drugs, a small percentage of which become commercial products. Thus, it is no surprise that rates of return on the drugs which *do* succeed are well above pharmaceutical companies' costs of capital. The return on the successes has to pay for the research and development on the failures. Oil companies have to pay for dry holes from the profits on successful wells.

---

[15] See Appendix B for a description of Gas Inventory Charges.

Computer software companies have to pay for programs that fail with the profits on the ones that succeed.

CONCLUSION

These three economic concepts -- the difference between regulated and competitive prices, the fact that only nondiversifiable risks affect the cost of capital, and the need for compensation beyond the cost of capital for negatively asymmetric outcomes -- provide the tools necessary to analyze the risks of the pipeline industry. That is the topic of the following chapters.

Chapter 5
Risk of the Interstate Natural Gas
Pipeline Industry: Summary

The regulated gas utilities [both pipelines and gas distribution companies] of the Eighties have stood out as
being especially vulnerable to business risk -- perhaps
more than they were during the supply-short period of
the Seventies. In the current decade, fuel oil competition has put this industry group through its paces as
never before (*Value Line* 1985, 465).

Our analysis of regulatory risk for gas pipelines began when the Interstate Natural Gas Association of America (INGAA) commissioned the
authors[1] to study the risk of the interstate natural gas pipeline industry
(hereafter pipeline industry).[2] INGAA was concerned that there might
be fundamental problems with the current system of pipeline regulation.
Most particularly for this study, INGAA was concerned that there
seemed to be a difference between regulators' perception of the industry's risks and the risks perceived by the pipeline companies themselves.

In the course of the work, we spoke to many people familiar with the
industry, including pipeline company staff, former regulators, and outside observers. We reviewed published works about the industry, court
decisions, and a number of decisions and orders of the industry's chief
regulatory body, the Federal Energy Regulatory Commission. We examined data supplied by INGAA and gathered additional data on our own.
We also considered certain important, general features of the economics
of rate regulation that are not always recognized by either regulators or
the companies they regulate. The process has led us to certain conclusions about the risks facing the industry and the more general issue of
whether the current regulatory system is sustainable in the long run.

---

[1]   The findings in this and subsequent chapters were based on A. Lawrence Kolbe,
William B. Tye and Stewart C. Myers (1991).

[2]   Broadly speaking, the natural gas industry has three components: production,
transmission, and distribution. Interstate natural gas pipelines occupy the middle
component. As discussed more fully in Appendix B, interstate pipelines traditionally bought gas from producers and sold a bundled service of gas and gas transportation to either distribution companies or end-use customers.

Our analysis uncovered serious problems facing the pipeline industry and its regulators. However, the study is diagnostic and not a brief for treatment. We focused solely on risk and not on other goals and concerns that should affect regulatory policy.

Moreover, we do not assume that risk is bad. Competitive businesses do not always 'solve' risks by reducing or eliminating them. More competition would pose new problems for gas pipelines only if they were asked to bear transition costs without adequate compensation, or if regulation impeded their ability to compete and earn a competitive profit.

Our main findings are:

1.   Natural gas pipelines are *not* low-risk businesses. They are not traditional public utilities with protected markets and stable and adequate returns. In some dimensions -- exposure to regulatory rule changes, for example -- they are definitely high-risk businesses. The view that the pipeline industry's troubles are past and that regulation can be counted on to fix any remaining problems is simply not borne out by the evidence.[3]

2.   The pipeline industry has endured roughly a decade of extreme danger and difficulty. Many pipelines suffered enormous losses. The specific risks that caused these losses seem to be receding. However, new risks threaten, reflecting unsettled markets, unsettled regulation, new and more intense competition, and more difficult environmental and operating requirements. Some pipelines may face less risk now than in the 1980s, but pipelines are generally much riskier than in the 1970s.

---

[3]   Within months of the writing of these conclusions, observers were shocked to learn of the bankruptcy of one of the largest interstate pipelines, Columbia Gas System, Inc. (*Wall Street Journal* 1991b, A3).

3.   Pipelines' business strategies and profits will be constrained by competition and also by traditional rate of return regulation. Competition is increasing and rate of return regulation is getting tighter. However, regulators (and the courts) have not completely specified the regulatory system or the boundaries of competition. Pipelines must plan future investments without knowing the detailed rules of the game and under a cloud of doubt whether any such hybrid system of regulation and competition can be sustained in the long run.

4.   One thing is nevertheless certain: if competition creates downside risks, and tight rate of return regulation eliminates profits above the cost of capital, then pipelines cannot earn fair profits on average. Consequently, investment in the natural gas pipeline industry will eventually be retarded and biased toward safe activities. Risky investments necessary to provide reliable service will be discouraged.

BACKGROUND

The Gas Pipeline Industry in the 1970s and 1980s

Most pipelines in the 1970s were traditional regulated utilities, selling an artificially cheap product into markets with ample demand. A perceived future shortage of natural gas led to aggressive bidding by pipelines to lock in supplies through long-term take-or-pay contracts.

But energy prices fell in the 1980s, and the shortage of natural gas turned into a 'bubble' of excess supply. New gas could be obtained at much lower prices than before.

The Federal Energy Regulatory Commission (FERC) decided to give local distribution companies and their customers access to this low-priced gas. Local distribution companies were relieved of minimum bills (contractual obligations to pay for natural gas from pipelines). Pipelines, however, were left with high-priced take-or-pay contracts with producers, and were therefore forced to renegotiate (i.e., buy out) these contracts. The costs of these buy-outs was about $9 billion, of which pipelines absorbed about $3.4 billion. Many pipelines suffered enormous losses, and on average their earned returns were far below both their allowed returns and local distribution companies' returns during the mid- and late 1980s.

Stockholders in natural gas pipelines naturally suffered. As shown in Figure 5-1 they earned far less than market averages in the 1980s, and were even further below the predicted return given their risk. On average, pipeline stockholders would have been much better off investing in Treasury bills.[4]

---

[4]   A dollar invested in Treasury bills in 1980 would have been worth $2.27 in 1990, compared to $1.51 for pipelines.

Figure 5-1
Predicted vs. Actual Returns for 4 `Pure Play' Gas Pipelines

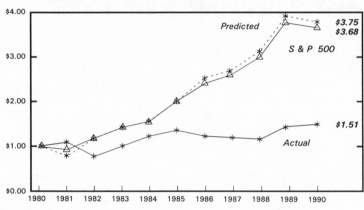

Source: Value Line, Merrill Lynch, Ibbotson Associates, Inc.

The regulatory changes that relieved pipeline customers of minimum bill requirements also gave 'open access' to gas suppliers, users, or brokers. That put pipelines in two businesses: the traditional merchant business of buying, shipping and reselling gas, and a new transportation business, in which pipelines simply charge a tariff for shipping gas owned by others. Transportation volumes have been growing steadily as a fraction of total throughput.

The Gas Pipeline Industry in 1991

The structure of regulation and competition in the 1990s is still unclear. Nevertheless, the following rough snapshot of the 'typical' gas pipeline in 1992 is a useful starting point. The pipeline must deal with four major groups: suppliers (gas producers), marketers (buyers and sellers of gas), regulators (primarily FERC), and customers (generally local distribution companies). Think of what each group wants.

Gas producers want to sell available production at the highest possible price. They will force gas pipelines to honor existing take-or-pay contracts or will extract as much as possible from pipelines that renegotiate.[5] At the same time FERC gives them the initiative to propose higher prices on certain existing contracts and to walk away from the contracts if pipelines do not accept the higher prices. If necessary, they will bypass the pipeline's merchant function and sell directly to local distribution companies, in this case paying for transportation only.

Local distribution companies want adequate supplies at the lowest possible prices. They can keep the right *to demand* gas, and to take advantage of the pipeline's obligation to serve, but try to beat the pipeline's price by dealing directly with producers, for example, by buying gas in the spot market and paying the pipeline for transportation only. Thus gas pipelines may be forced to supply purchased gas at times of peak demand when the spot price is high. At off-peak, the pipeline may be left with unsold gas.

Marketers buy and resell gas whenever they see a profit. They want open access to as many potential customers as possible.

FERC wants to encourage competition, thus placing more responsibility (and risk) on pipelines' shoulders. At the same time, it wants to maintain rate of return regulation to prevent 'excessive' profits and keep prices at the burnertip down.

Now consider the gas pipeline's alternatives. It is locked into a declining, but still substantial obligation to buy, ship, and resell gas -- i.e., a substantial merchant business. It must honor (mostly renegotiated) take-or-pay contracts (some of which have years still to run) but will have no special access to low-priced gas. It must make transportation available to customers who buy in spot markets. At the same time it must keep capacity available and have gas ready to sell to local distribution companies at peak periods or when spot prices are high. Off-peak, or when spot prices are low, it often has difficulty selling all of the gas it has contracted to buy. It cannot get out of the merchant business because of

---

[5]   Most take-or-pay contracts have been or are being renegotiated. However, renegotiated contracts may still offer favorable terms or prices to producers.

obligations to local distribution companies, yet its merchant volume is heavy only when customers and producers want that volume to be heavy.

Demand for the pipeline's *transportation* business is also uncertain. Here again the pipeline faces a difficult asymmetry. Regulation prevents it from raising tariffs when demand is high, but discounting is allowed when demand is slack. The volume of discounted shipments has been increasing rapidly.

There are obvious downsides for the pipeline but no evident upsides. FERC regulations do not allow it to raise gas prices or tariffs for transportation when demand is strong. It may earn less than the rate of return allowed by regulators, but if regulation is tight it will have a hard time earning more.

Of course, regulation is not all negative. For example, FERC has provided for a 'gas inventory charge' or GIC designed to compensate pipelines who must (1) stand by to sell gas to customers (at predetermined prices) when it is in the *customers'* interest to buy, but (2) allow customers the option to transport third-party gas whenever that is cheaper for them. Obviously the customer gets the best of both worlds, and the pipeline the worst. The gas inventory charge is designed to even up the bargain. However, GICs are still experimental, and even those that are in place may well be rendered obsolete by 'comparability.'

Looking Ahead to Comparability

The foregoing snapshot of the gas pipeline industry in 1992 is not a stable picture. FERC has recently embarked on a fast track to full comparability, i.e., to a regulatory framework that allows producers or other third parties to compete across the board for gas pipelines' traditional customers. The new framework is to be implemented in proceedings launched by FERC's latest Notice of Proposed Rulemaking (NOPR) and as implemented in Order No. 636.[6]

---

[6]   *In Re Pipeline Service Obligations and Revisions to Regulations Governing Self-Implementing Transportation Under Part 284 of the Commission's Regulations,* Docket No. RM91-11-000, issued July 31, 1991. This proceeding is frequently

The existing regulatory framework has moved part of the way to comparability and open access. Until the 1980s, pipelines owned virtually all of the gas they shipped. Producers or other third parties now offer interruptible service to gas customers, and customers also buy gas from third parties for use with firm pipeline transportation service. The remaining volumes (roughly 20 percent of gas delivered to the final market) are pipeline firm sales service, i.e., the pipeline provides the gas supplies and capacity to deliver (up to) an agreed amount of gas. At present, gas pipelines retain an advantage in providing firm sales service in winter. For example, a potential competitor would not have direct access to the storage facilities necessary to assure firm service in peak demand periods.

Under full comparability, storage and other pipeline services would be unbundled and available to potential competitors. Thus, a producer with access to low-cost gas could sign a firm supply contract with a local distribution company, buy transportation capacity, storage, etc., from the gas pipeline, and thus displace the pipeline as a seller of gas. (The pipeline would continue, of course, as a regulated carrier.)

The proposed rule is only in draft, and it will undoubtedly be revised before final issue. Many details remain to be worked out. Thus, it cannot be analyzed definitively here. Still, there are four observations we can make.

1.   Some pipelines may thrive in competitive gas sales markets. Others, caught with long-term take-or-pay contracts renegotiated to cover existing merchant business, are at risk of a replay of the 1980s. On average pipelines are likely to lose under comparability, because they already hold the remaining firm sales market, and the

---

called the 'mega-NOPR.' For the final rule, see FERC, *Order No. 636, Final Rule,* Docket No. RM91-11-00, *et al.,* 59 FERC ¶61,030. The text of this book was completed prior to the issuance of the final rule. As written it appears to address and account for many of the issues of asymmetry that we identify. The final outcome, of course, will depend on court challenges and actual implementation of the policies.

more profitable parts of this business will be the first targets of their new competitors.

2. Gas pipelines' ability to recover the costs of further contract restructurings is unsure. This is a major risk from comparability. The mega-NOPR suggests the Commission *intends* to allow pipelines to recover a substantial fraction of such costs, but competition in the post-comparability marketplace could easily render this promise empty. Moreover, since pipelines have not received compensation for the risk that they might have to bear comparability transition costs, *any* share of the transition costs is too large a share.[7]

3. Some features of the mega-NOPR may be attractive to pipelines. For example, the proposal to shift to a straight fixed-variable rate design may reduce pipeline risks.[8] Of course, this helps pipelines only if the allowed rate of return is not reduced correspondingly.[9]

4. A host of regulatory and operational details remain to be worked out. Will profits or losses on pipelines' sales business be set off against

---

[7] Gas Inventory Charges (GICs) were intended to compensate for the risks of signing contracts to meet firm sales commitments, but GICs at present are ill-defined and not widely implemented. Existing GICs in any case were not designed to compensate for costs of transition to comparability.

[8] Currently, most pipelines have to recover some of their fixed costs from charges for the gas itself, rather than in annual demand charges. The mega-NOPR apparently would shift more of the fixed costs to demand charges, although details are not clear.

[9] FERC staff have proposed reduced rates of return on equity for straight fixed-variable rate structures, but changes in rate design historically have been responses to other major changes. The rate of return should not be reduced if other changes offset the effect of a fixed-variable rate design.

allowed returns in transportation? How will
storage and other unbundled pipeline services be
priced and/or allocated among the pipeline and
other users -- users which in most cases will be
the pipelines' competitors in the end-of-pipe
market for natural gas? How can transportation
systems best be operated under the new rules,
and who is at risk for the costs and inefficiencies
that will inevitably arise until efficient operating
procedures are worked out?

The NOPR to implement full comparability opens up dozens of rules and
procedures for argument and maneuver by all the interested parties.
Standing in 1992 there is no way of saying which constituencies FERC
will bless.

RISKS FACING NATURAL GAS PIPELINES FOR THE 1990s

'Risk' has several meanings. Pipelines could be high-risk in one sense
of the word and average risk in another.

Economists and financial analysts (who usually testify about fair rates of
return) think of risk as the overall volatility or unpredictability of earnings or of the return to investors. Risk is often expressed as a statistical
measure. The level of risk determines the expected rate of return demanded by investors and thus determines the cost of capital.

Managers usually think of risks (plural) as *things that can go wrong*, as
possible adverse events that could set back earnings or reduce asset
values. Their first concern is not with the cost of capital, but with
whether forecasted returns can actually be achieved. Managers often
have difficulty condensing a list of risks into a numerical estimate of the
cost of capital or fair rate of return. Economists and regulators, on the
other hand, often forget to ask whether the cost of capital can actually be
earned.

In this instance it is best to concentrate first on risks as managers of gas
pipelines see them. A manager looking ahead from 1992 would see
many possible adverse outcomes. A partial list follows.

Unsettled Markets

Patterns of supply and use are changing. Fixed pipeline capacity may be left underutilized. Pipelines may be forced to discount purchased gas to sell it, or discount tariffs to attract transportation business. There is increased exposure to fluctuations in world energy prices. If energy prices decline, it may be impossible to market gas at prices which recover contract costs and earn a fair rate of profit.

Unsettled Regulation

FERC has for several years been searching for mechanisms to achieve the benefits of competition for pipeline customers within the legal structure of regulation. The search has generated a bewildering series of rulings, several of which were sent back to FERC by the courts. The mega-NOPR to establish full comparability is just beginning, and other issues remain unresolved. Thus the rules of the game are still not clear. For example:

1. If full comparability is achieved, will pipelines be able to recover the additional costs associated with further contract restructurings, or will allowed rates of return be increased to compensate? Even if FERC regulations permit such recovery, will the market?

2. If full comparability is not achieved, or is instituted with limits on the amount pipelines can charge for gas, how will gas inventory charges be implemented? Will they generate adequate premiums for pipelines that guarantee supplies to local distribution companies?

3. Will pipelines be allowed to increase allowed tariffs to fully offset revenues lost from discounting?

4.   Will unbundling allow further cream-skimming
        by customers? That is, will separate billing of
        storage and other services and costs allow cus-
        tomers to pick and choose and improve their
        bargaining position with pipelines?

Of course FERC orders will still be subject to judicial approval or remand. Moreover, even when (or if) the rules of the game settle down,
there will still be lingering doubt about whether the rules will stand if
market conditions shift dramatically. 'Mistakes' are inevitable in a risky
industry, and there will surely be pressure on legislators and regulators
to protect constituencies from painful outcomes. Gas pipeline investors
and managers will remember that they were not a favored constituency
in the 1980s.

Competition

Competition among pipelines has obviously been increasing. Discounting
has grown rapidly. At the same time, suppliers and customers can take
end runs around pipelines on sales, while pipelines remain the suppliers
of last resort. Pipelines are concerned with their ability to meet peak
demands in a world where transportation customers can shift on short
notice into merchant customers.

Comparability and Open Access

The move to full comparability will expose pipelines to competition in
firm sales of natural gas. That is not a bad thing; competitive risks are
good risks. However, pipeline managers worry about the transition to
comparability. For example, most pipelines renegotiated take-or-pay
contracts with producers but did not eliminate them. The new contracts
typically obligate pipelines to take (or pay for) gas supplies adequate for
their obligation to serve local distribution companies. The existing regulatory framework limited access to the market for firm gas sales and
provided gas pipelines a reasonably stable and assured market. Full
comparability would remove that stability and assurance without removing the long-term contracts or providing full recovery of contract renegotiation costs.

Pipeline managers also worry that regulation will deny them a fair chance to earn a competitive rate of return. For example, good performance in the market for sales of natural gas could be offset by reductions in allowed profits in transportation. Also, the technical right to impose a gas price surcharge for transition costs is empty if competitors can undercut that price.

Environmental and Operating Risks

Operating a pipeline is more complicated than it used to be. Two businesses, merchant and transportation, must be run through the same pipe. Balancing many different customers' inputs and deliveries has become much more difficult.

Certain traditional operating risks remain, for example, extremely cold weather which could cause equipment to freeze, curtailing throughput and deliveries. There are also newly recognized risks, e.g., the possible costs of cleaning up PCBs or other environmental hazards.

Conclusions

The risks we have just reviewed fall into two categories:

1. Possible major new risks as a result of full comparability as defined in the mega-NOPR. The risks in this area may prove massive, but the mega-NOPR is in such an early stage that it is impossible to determine just what its impact will be.

2. The current day-to-day risks of the gas pipeline business, including unsettled markets, unsettled regulation, competition, and environmental and operating risks. None of these latter risks, taken alone, is as serious as the risks created by high-priced take-or-pay liabilities in the 1980s. The risks of the 1980s could have been fatal. In the 1990s pipelines do not face one or two massive

risks, but a much longer list of (still material) risks in their core business.

Are the natural gas pipelines riskier now than in the 1980s, the mega-NOPR aside? Is it better to face one or two massive risks or, say, a dozen lesser, but still important, risks? There is no mechanical answer, but one can die from a thousand small cuts as surely as from one stroke of the sword. It is clear that pipeline managers face more complex prospects than before. It is also clear that pipelines are no longer safe businesses protected by regulation. They may have been so in the 1970s, but not now.

It is tempting to think that gas pipelines would become safe if they could shed their merchant obligations completely and operate as common carriers, offering transportation only. This is fallacious; the intrinsic risk of the pipeline transportation business is substantial. For example, oil pipelines, which have been common carriers all along, have on average suffered much more volatility in earnings than gas pipelines from 1977 to 1988.[10,11]

REGULATION AND COMPETITION

Rate of return regulation is supposed to allow a fair chance to earn a return commensurate with the risks of the business. If the prospective return is adequate, then gas pipelines' new risks should not pose any new policy concerns; pipelines should trust the regulatory process.

In fact, pipelines do not completely trust the process. They perceive *regulatory* risks. This is not a lack of trust in regulators as individuals, but a lack of confidence in the process and structure of regulation. Some

---

[10]   Note that this period *includes* gas pipelines' time of troubles in the mid-1980s. For 1977-1984, the volatility of oil pipelines' earnings was dramatically higher than for gas pipelines. See Figures 8-11, 8-12, and 8-13.

[11]   Large investments with few alternative uses are intrinsically risky. Demand charges partially mitigate this risk for gas pipelines, as 'throughput and deficiency agreements' do for the reported earnings of oil pipelines. But the level of demand charges is itself uncertain over the life of a gas pipeline.

of this lack of confidence is based on painful memories from the 1980s, particularly of the repeal of customers' minimum bill provisions without offsetting changes in take-or-pay obligations to gas suppliers. Pipelines also remember that the rates of return allowed by FERC were *not* increased in the 1980s (see Figure 8-3), when risk (in the business manager's sense of the term) increased in an obvious and dramatic way. They remember the years of confusion as FERC orders bounced back and forth between the Commission and the courts. Looking forward, with the mega-NOPR under way, the rules of the game are still unclear. The only certainty is that competition will increase.

FERC has encouraged competition and removed constraints in market structure and contractual arrangements. Congress will have completely deregulated all gas production by 1993. Pipelines continue to operate as merchants of gas, but now also sell stand-alone transportation service. Local distribution companies can buy directly from gas producers in spot or long-term markets for gas. The market for firm gas sales is now also to be opened up to competition. However, many gas pipelines will be left with obligations to producers and local distribution companies, and all will remain subject to rate of return regulation.

Pipeline managers believe regulation of transportation is becoming increasingly tight. FERC staff negotiates aggressively to hold down pipelines' profits, for example, by using optimistic throughput forecasts to calculate allowed tariffs. The FERC staff has also argued that discounting be ignored in setting transportation tariffs, thus forcing pipelines to 'eat' revenue losses from discounting.[12] Yet everyone must now understand that discounting is a fact of life.

The FERC now seems reluctant to use the certification process to protect existing pipelines from new entry and new pipelines from mistaken entry. Pipelines are at risk if they misestimate required capacity, yet there is no policy to compensate for such risks.[13] Also, the Commission may not

---

[12] We understand that some FERC staff have taken this position at least through early 1991.

[13] If pipelines expect to earn only their cost of capital if capacity does not prove excessive, but less than their cost of capital if capacity does prove excessive, on average they will expect to earn less than their cost of capital.

rule on whether a new facility will be treated on an incremental cost or 'rolled-in' cost basis until after it is built, creating the risk that incremental cost pricing will lead to excess capacity.

Immediate Problems

The resulting hybrid system of competition and regulation poses both immediate and longer-run, structural problems for gas pipelines. Consider the immediate problems first.

1. Deregulation often triggers a period of confusion and experimentation as companies try to develop business strategies to cope with new markets and new competition. Then there is a shakeout as it becomes clear which strategies will succeed. This process has played out in the domestic airline industry and is under way in the commercial banking industry. Gas pipelines are in the early stages of the same process.

2. Regulation has not settled down. The rules of the game are not clear.

3. If there are new risks or setbacks to the natural gas industry as a whole (producers, pipelines, and local distribution companies taken together), any imbalance between risk and reward will probably fall disproportionately on pipelines. This is exactly what happened in the 1980s.

4. Competition will limit regulators as well as pipelines. Regulators may not be willing or able to put enough 'slack' in the ratesetting process to give pipelines some upside to offset the risks they face.

5. Traditional methods of estimating costs of capital do not yield *allowed* rates of return sufficient to offset the risks pipelines face. DCF methods do

not automatically adjust for these risks, for
example (see Chapter 3).

Of course, many risks facing pipelines are diversifiable and should not
affect the cost of capital. But that is not the point: if pipelines can
expect to earn only their allowed rates of return when things go well,
then the allowed rates of return must exceed the cost of capital if things
can go badly. Otherwise, pipelines cannot earn the cost of capital on
average.[14]

Structural Problems

Any attempt to combine active competition and tight rate of return regulation will generate conflicts and inconsistencies. First, the time patterns
of competitive returns to capital do not match patterns assumed in regulation. Second, competition creates downside risks, but regulation limits
upside potential.

A business subject to competitive as well as regulatory discipline gets the
worst of both worlds. This is a longer-run, *structural* risk facing gas
pipelines. This structural risk shows up in two major ways.

1.   Capital charges do not match over time. Capital
      recovery may be impossible for firms facing
      both regulation and competition.

2.   Regulation creates asymmetric risk. The upside
      potential created by competition is cut off, but
      the downside risk remains. In other words,
      competition offers both a carrot and a stick;
      adding tight rate of return regulation takes away
      the carrot.

---

[14]   Pipeline take-or-pay losses in the 1980s offer clear examples. Pipelines sold gas at
        cost, earning no profits on sales. Yet they suffered large losses when their mini-
        mum bill sales provisions were abrogated while take-or-pay obligations to producers
        remained in force. There was downside risk without upside potential. Companies
        cannot expect to earn a fair return on average under such conditions. Pipelines may
        face the same 'no-win' prospect in the transition to comparability.

Capital Charges Do Not Match Over Time. Figure 4-4 shows why companies facing both regulation and competition often get the worst of both worlds. Regulation keeps capital charges (the sum of allowed earnings and depreciation) on a steadily declining path. Competition forces low *initial* capital charges but allows increases in line with inflation. Thus regulation 'allows' artificially high capital charges on new assets, and artificially low capital charges on old assets.

If competition denies the early rewards of regulated business, but regulation denies the increases that naturally come with inflation, there is no way to earn a fair and reasonable return.

Asymmetric Risk. Suppose FERC imposes tight rate of return regulation that minimizes pipelines' chances of earning more than the allowed return. Suppose further that the allowed rate of return is the true cost of capital, but that competition may prevent the pipeline from earning the allowed rate of return at some times or in some circumstances. Then the expected (average) return *must* fall short of the cost of capital. The cost of capital is *defined* as an expected rate of return, i.e., the rate of return averaged over possible successes and failures. If failure means earnings which fall short of the cost of capital, then returns must exceed the cost of capital when failure is avoided. In other words, regulators must provide some 'upside' -- an allowed rate of return above the cost of capital -- to compensate for the downside created by competition or by the various risks facing gas pipelines.

If regulators provide no upside, but allow only what investors require on average (the cost of capital), they will get inadequate or misplaced new investment, and risky, aggressive investment will be chilled.

These structural problems are created by the attempt to combine traditional regulation with active competition in a risky industry. The timing and form of these effects are harder to pin down than the specific risks noted earlier. There is considerable momentum in corporate investment programs, and there may still be 'pockets' of safe opportunity, stable markets protected by regulation or location, where expansion is attractive despite unsettled regulation.

Thus these structural problems may not show up in any immediate or dramatic way. The regulatory system is sustainable in the short run, barring drastic changes in energy prices or patterns of gas supply and use. In the long run, however, new investment will probably disappoint regulators and policy makers unless FERC loosens or modifies the regulatory rules of the game.

PROSPECTS

Gas pipelines in the 1990s will not be low-risk businesses. They face many risks, both specific and structural.

No one specific risk seems as dangerous as the massive risks faced in the 1980s, in particular the obligation to deal with high-priced take-or-pay contracts with producers without the hedge of customers' commitments to minimum bills, although contract restructuring under comparability could replay that risk. Even without the transition to comparability, the cumulative effect of the risks of the 1990s is considerable.

The structural problems are deeper and in the long run may be more serious. If competition is encouraged, but tight regulation 'caps' earnings at the cost of capital, then investment will be inhibited and biased toward safe ventures.

POLICY ISSUES

The risks facing gas pipelines in the 1990s have implications that reach beyond the parochial interests of the pipelines and their customers. Obviously it is important for regulators to give gas pipelines and their investors the opportunity to earn a fair rate of return on average. It is equally important to encourage competition and efficiency, and to avoid artificial constraints on pipeline industry structure. But there is a tension between traditional regulation, on the one hand, and competition and efficiency on the other. Inappropriate regulation creates a danger for the overall economy, the risk that badly needed additions to pipeline capacity will be warped or stunted.

Solutions to these problems require consideration of more than just pipe-line risk, and so are beyond the scope of our study. However, we can say that 'more of the same' is unlikely to do the job. For example, an allowed rate of return above the cost of capital, to compensate for asymmetric risk, can encourage sensible risk-taking by pipeline management. But this approach will not be effective where competition prevents pipelines from earning the higher allowed rate. In this and other cases only structural change in cost-of-service regulation will create economically sensible incentives.

The combination of competition plus tight regulation with an allowed rate of return equal only to the cost of capital will not work. Policy makers must address fundamental structural problems to create a framework for the health and growth of the gas pipeline industry.

Chapter 6
Recent Trends in the
Interstate Gas Pipeline Industry

> Everybody says we are at war now, only it has not been
> announced. We think we are very safe here in the is-
> lands, in fact, this place is better defended than the U.S.
> With all the fleet here, and all the army defenses, I don't
> think anybody will harm Hawaii. Japan couldn't do
> anything, because they have no base near here, and they
> couldn't do much 6,000 miles off. (Private correspon-
> dence to Adrienne Hawkins Tye, date-line, Honolulu,
> May 22, 1941)

BACKGROUND

By any standard, the pipeline industry has faced a period of upheaval
during the 1980s, but debate has arisen over the level of risk the industry
faces today. Traditionally, a pipeline served a bundled merchant and
transportation function. The pipeline would buy gas from producers 'at
the wellhead' under contract, transport it, and sell the gas 'at the city
gate' to distribution companies or end-use customers. The ratemaking
process that went with this environment is depicted in Figure 6-1. The
return to invested capital is partly depreciation (return *of* capital) and
partly operating earnings (return *on* capital). The cost of providing
service, i.e., the return to capital plus operating expenses and the cost of
the gas itself, must be recovered from the pipeline's customers. This
cost can be viewed as having two parts: costs that are fixed in the short
run and costs that are variable in the short run.

Overall costs similarly have been recovered in a two-part rate design.
'Demand charges' are independent of throughput and give the customer
certain rights to capacity during peak periods. 'Commodity charges' are
paid for each unit of gas customers actually take. For reasons that have
changed over the years, fixed costs have been recovered partly in the
demand charge and partly in the commodity charge. The formula that
allocates fixed costs between demand and commodity charges has itself
changed over the years, as a matter of policy and with changing industry
economics.

Figure 6-1
The Traditional Natural Gas Ratemaking Process

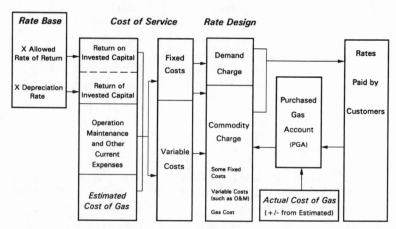

Source: Prepared testimony of Charles J. Cicchetti, Jr., Ph.D., before the Arkansas Public Service Commission in the matter of Investigation in the overall Gas Practices of Arkla, Inc., Docket No. 90-036-U, August 31, 1990.

The building blocks of the traditional ratemaking approach remain, but the process itself has become more complicated in recent years. One reason is the growing importance of stand-alone transportation service, as opposed to bundled transportation and sales service. Another is the greater complexity of the way fixed costs are allocated to various services. Figure 6-2 depicts a hypothetical rate design today. The complexity is immediately evident.[1]

Beginning in the 1950s, following the pivotal *Phillips* decision,[2] the price of gas shipped over interstate pipelines was regulated at the wellhead by the Federal Power Commission and its successor, the Federal Energy Regulatory Commission (FERC). By the 1970s, the result was a shortage of gas in interstate commerce, as demand exceeded supply at the

---

[1]   Even Figure 6-2 is overly simplified. It contains no provision for a 'Gas Inventory Charge' or distance-sensitive rates, for example.

[2]   *Phillips Petroleum Co.* v. *Wisconsin*, 347 U.S. 672 (1954).

Figure 6-2
Hypothetical Cost of Service

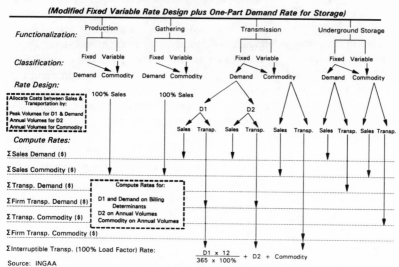

(Modified Fixed Variable Rate Design plus One-Part Demand Rate for Storage)

Source: INGAA

artificially low, regulated price. To remedy the shortage of natural gas, the Natural Gas Policy Act (NGPA) of 1978 set the stage for a gradual decontrol of prices during the late 1970s and the 1980s. The NGPA decontrol mechanism permitted higher ceiling prices on new gas discoveries than on gas from existing fields.

The initial high ceiling prices on 'new' gas under the NGPA coincided with very high energy prices in 1979 and the early 1980s. Moreover, at the time, these prices were widely expected to rise even further. 'Old' gas under the NGPA was still subject to low ceiling prices, and pipelines charged a 'rolled-in' price for gas: an average of the price for 'old' and 'new' gas. The result was that pipelines signed long-term contracts for new supplies at what turned out to be very high prices indeed. These price increases were justified by the need to obtain new supply to satisfy what had been a chronic shortage and by the expectation at the time the contracts were signed that eventual price increases could be recovered on a rolled-in basis.

Instead, the high prices early in the decade gave way to a fall in energy prices and a glut of energy supplies. This created a mismatch between high prices under these new long-term contracts and low spot market prices for gas. This mismatch led to economic hardship that someone -- producers, pipelines, distributors, or final customers -- was going to have to bear and to demands for fundamental changes in the regulatory system. Following a series of decisions by the FERC (especially Order 380[3] in 1984, which relieved pipeline customers from 'minimum bill' obligations to pipelines without relieving pipelines of 'take-or-pay' obligations to gas producers[4]) pipelines faced insolvency-threatening losses of billions of dollars in unrecoverable expenditures for gas that could not be sold at the contract price.

At the same time, and partly in response to court-mandated concern over the take-or-pay problems facing the pipelines, the FERC was rewriting the rules by which it regulates gas pipelines. In particular, Order 436 and its successor following remand, Order 500,[5] encouraged pipelines to

---

[3]   Order No. 380, *Elimination of Variable Costs From Certain Natural Gas Pipeline Minimum Commodity Bill Provisions*, 49 Fed. Reg. 22,778 (June 1, 1984), FERC Stats. & Regs. ¶30,571, *reh'g denied and stay granted in part*, Order No. 380-A, 49 Fed. Reg. 31,259 (Aug. 6, 1984), FERC Stats. & Regs. 1982-1985 ¶30,607, *reh'g denied*, Order No. 380-D, 29 FERC ¶61,332 (1984), *aff'd in part, remanded in part sub nom. Wisconsin Gas Co. v. FERC*, 770 F.2d 1144 (D.C. Cir. 1985), *cert. denied sub nom. Transwestern Pipeline Co. v. FERC*, 476 U.S. 1114 (1986), *order on remand.* Order No. 380-E, 35 FERC ¶61,190 (1987).

[4]   Under minimum-bill provisions in approved FERC tariffs, a distributor had to pay for a minimum amount of gas whether or not the distributor could sell that gas. Under take-or-pay provisions in pipeline contracts with producers, pipelines similarly had to pay for a minimum proportion of the gas covered by the contract whether or not they accepted it (generally with five-year makeup periods).

[5]   Order No. 436, *Regulation of Natural Gas Pipelines After Partial Wellhead Decontrol*, 50 Fed. Reg. 42,408 (October 18, 1985), FERC Stats. & Regs. ¶30,665 (Oct. 9, 1985), *modified*, Order No. 436-A, 50 Fed. Reg. 52,217 (Dec. 23, 1985), FERC Stats. & Regs. ¶30,675 (Dec. 12, 1985), *modified further*, Order No. 436-B, 51 Fed. Reg. 6398 (Feb. 14, 1986), *reh'g denied*, Order No. 436-C, 34 FERC ¶61,404 (Mar. 28, 1986), *reh'g denied*, Order No. 436-D, 34 FERC ¶61,405 (Mar. 28, 1986), *reconsideration denied*, Order No. 436-E, 34 FERC ¶61,403 (Mar. 28, 1986), *vacated and remanded, Associated Gas Distributors v. FERC*, 824 F.2d 981 (D.C. Cir. 1987). Order No. 500, *Regulation of Natural Gas Pipelines*

adopt 'open access,' i.e., to permit final customers and third parties to transport their own gas on the pipeline system. Order 436 effected no relief for pipelines' take-or-pay liabilities, and this feature led to remand and to Order 500 and its amendments. Under current regulations, but with further court reviews likely, pipelines have a mechanism to recover only a part of their prudently incurred take-or-pay costs from final customers. The losses pipelines have borne from take-or-pay disputes remain large, and some of these disputes are ongoing.

The FERC's aim during much of the 1980s appears to have been to find a way to achieve the benefits of competition for pipeline customers, within the legal structure of regulation. The resulting new rules in turn frequently have been overturned on judicial review, which has led to still newer versions as the FERC attempts to achieve a legally permissible system that reconciles competition and regulation in a period when energy markets are subject to dramatic and unpredictable change. As this book is written, a number of important FERC rulings either are under legal challenge, have been remanded, are not fully operational, or are the subject of two or more of these uncertainties.[6]

After Partial Wellhead Decontrol, 52 Fed. Reg. ¶30,334 (Aug. 14, 1987), FERC Stats. & Regs. ¶30,761, *extension granted*, Order No. 500-A, FERC Stats. & Regs. ¶30,770, *modified* Order No. 500-B, FERC Stats. & Regs. ¶30,772, *modified further*, Order No. 500-C, FERC Stats. & Regs. ¶30,786 (1987), *modified further*, Order No. 500-D, FERC Stats. & Regs. ¶30,800, *reh'g denied*, Order No. 500-E, 43 FERC ¶61,234, *modified further*, Order No. 500-F, FERC Stats. & Regs. ¶30,841 (1988), *reh'g denied*, Order No. 500-G, 46 FERC ¶61,148 (1989), *vacated and remanded, American Gas Association v. FERC*, 888 F.2d 136 (D.C. Cir. 1989) (AGA-I). Order No. 500-H, Final Rule, FERC Stats. & Regs. ¶30,867 (Dec. 13, 1989). Order No. 500-I, FERC Stats. & Regs. ¶30,880 (Feb. 12, 1990), *aff'd in part and remanded in part, American Gas Association v. FERC*, 912 F.2d 1496 (D.C. Cir. 1990) (AGA-II), *cert denied sub nom. City of Willcox v. FERC*, 59 U.S.C.W. 3562 (U.S. Feb. 19, 1991), *Orders on remand*, Order No. 500-J, 54 FERC ¶61,148 (1991), Order No. 500-K, 55 FERC ¶61,020 (1991).

6 "Since January, 1988, the FERC has suffered more remands by the United States Circuit Courts of Appeals of its orders, in whole or part, than it has been affirmed." (Warzynski 1990, 269, footnote omitted). According to the American Bar Association's Public Utility Section *Newsletter* (1990, 7), "FERC's current record of remands and reversals by appellate courts is unprecedented in regulatory history."

The number of these rulings and the frequency of remand have created a bewildering array of changes and regulatory mechanisms that few outside the industry are ever likely to understand fully. Moreover, important details of application may show up either in individual company rate orders or in settlement agreements between pipelines, customers, and FERC staff. To further complicate matters, the rules the Commission promulgates are not always followed in proposed FERC staff settlements. The result is that even a reading of the major rulemakings may leave an outsider unclear about the actual economic impacts of the rules. Appendix B to this book provides a summary of what we believe are the most important developments in this history.

Since the bulk of this book was completed, FERC has issued its Notice of Proposed Rulemaking on 'comparability.'[7] Under comparability, pipelines would 'unbundle' and charge separately for the components of their firm sales service, most importantly storage.[8] Many important details of comparability remain to be decided, and this book therefore cannot assess the ultimate effects of comparability on pipeline risk. However, we do make a few observations on the impact of comparability below.

TWO VIEWS OF PIPELINE RISK

Perhaps unsurprisingly in view of this history, gas pipeline executives see the present as a time of tremendous risk as the industry adapts to a new environment. Competition is emerging as a major force, new operational problems are arising as pipelines learn to balance many uncoordinated users of what had been a tightly controlled transportation system, and the regulatory and legal standards governing pipeline tariffs and conditions of service are in flux.

---

[7]   *In Re Pipeline Service Obligation and Revisions to Regulations Governing Self-Implementing Transportation Under Part 284 of the Commission's Regulations,* Docket: No. RM91-11-000, issued July 31, 1991. This is often referred to as the 'mega-NOPR.'

[8]   The roughly 20 percent of final market delivered volumes that currently constitute pipeline firm sales are mostly made in winter, and storage is necessary to guarantee delivery on peak days.

Indeed, the pipeline industry is merely one of the regulated industries experiencing substantial regulatory reform. One of the central themes of these reforms is to create more incentive for management efficiency by injecting more competition into the process.[9]

These reforms represent a clear rejection of the older model of regulated monopoly. Under this concept, regulated firms were given a franchise and afforded protection from competitive entry and price competition. In return they provided reliable service, pursued conservative management styles, and perpetuated cross-subsidies that kept regulators and their constituencies happy. Since almost all prudent expenditures were billed through to customers, much of the risk of cost variances was borne by them. Customers, however, benefited to the extent that regulated firms had access to capital at low cost based on investor perceptions of low risk in a stable industry.

Regulatory reform and incentive regulation more specifically involve an explicit decision to put more of the risk/reward responsibility onto the shoulders of managers and their stockholders so that they will be motivated to achieve greater efficiencies. To the extent that regulatory reform succeeds, business persons and investors will perceive greater risk in these industries than before, by definition. Investors in these industries will necessarily have to be compensated for assuming these risks. Presumably the expected greater efficiencies achieved by shifting these risks from customers to regulated firms exceed the expected costs of greater offsetting compensation to investors for bearing those risks.[10]

---

[9] For a review of some of the common themes, see John R. Meyer and William B. Tye (1988).

[10] As an illustration, gas pipelines traditionally billed gas acquisition costs through to customers at cost, on the grounds that no risk implied no profit. The take-or-pay problems of the 1980s reveal that long-term contracts intrinsically are risky, since the hedge provided by minimum bills proved unsustainable in the face of dramatic price fluctuations. The FERC's 'Gas Inventory Charge' (GIC) tariffs were originally designed to permit pipelines to recover *future* prudently incurred take-or-pay costs, which would effectively restore minimum bills. However, it is hard to see why such GICs would be any more sustainable than minimum bills were in the face of future dramatic price changes. Thus GICs may instead evolve into 'insurance premiums,' to compensate pipelines for the risks incurred of maintaining a long-term supply portfolio without the corresponding obligation of customers to take that

Using this standard, the proposition that investor risks have increased in the gas pipeline industry as a result of regulatory reform is almost axiomatic. The question then becomes one of identifying the types of risks that have increased and the degree of increase in those risks.

Yet INGAA finds evidence of a contrary view among pipeline customers, regulators and regulatory commission staff, a view that risks in the gas pipeline industry are now rather low, relative to the past for the pipeline industry and relative to that of other industries. Our own conversations with former regulators confirm that this view exists. This alternative view points to the resolution, for good or ill, of most of the take-or-pay disputes with producers and to a series of FERC rulings designed to help pipelines adapt to today's market.

This disparity of views can have important consequences. If pipeline risks are truly low, an incorrect perception of high risk may lead to excessive charges to customers, as the FERC overestimates the rate of return it needs to allow to pipeline investors. Alternatively, if the FERC believes pipeline risks are low and they actually are high, the resulting substandard rates of return will lead to pipeline investments that are inadequate and/or of the wrong kind. That response may hinder or doom efforts by the industry and its regulators to adapt to new economic conditions. It also may retard the use of natural gas at a time when environmental and national security concerns make gas a particularly attractive fuel. Thus, an accurate picture of the level of risk the industry actually faces is vital to constructive regulation of the industry.

SUMMARY OF FINDINGS

Our analysis leads to findings in three areas:

        1.   The risks of regulated industries generally;

---

supply. This distinction has important implications both for the design of GICs and for pipeline risks, as discussed below.

Of course, comparability may change the picture considerably, by permitting pipelines to charge prices above gas costs and rendering GICs obsolete before they ever really are established.

2.   The risks of the pipeline industry specifically; and

3.   The long-run outlook for the current pipeline regulatory system.

Regulatory Risks in General

Two features of modern rate regulation create risks that may easily go unrecognized by regulators and regulated companies alike. First, regulation overcharges for capital assets early in their lives and undercharges for old assets, relative to competition. This difference has important implications for the risk of any regulated industry as competition becomes more important. Competition may prevent a regulated company's charging the high early rates that regulation permits on a new investment, while regulation may prevent the company's charging the high rates on old assets that competition permits.[11]   Moreover, regulated companies can have difficulty adapting to competitive markets, because different skills are needed to satisfy regulators and to respond to competition.

Second, some risks that do not affect the cost of capital may nonetheless affect the rate of return pipelines should be allowed. Common regulatory practice is to equate the allowed rate of return with the best estimate of the firm's cost of capital. However, it turns out that this practice may *not* provide investors a fair opportunity to earn the cost of capital. In particular, if the firm faces material downside risks without an offsetting upside opportunity, investors require risk compensation *in addition to* an allowed rate of return equal to the cost of capital.

The cost of capital is a break-even rate of return, and the average of breaking even and incurring a loss is a loss. The goal that truly permits companies to break even is an *overall* rate of return, on successful and unsuccessful investments *combined*, that investors expect will equal the

---

[11]   Also, the 'front-end load' in regulated prices distorts decisions by pipeline customers, and hence pipelines. These characteristics are economically undesirable even if growing competition were not an issue.

cost of capital. That goal cannot possibly be achieved by a policy that restricts the return on successful investments to the cost of capital, but permits returns well below that level in other circumstances. Just as oil companies need successful wells to pay for the dry holes, regulated companies need returns above the cost of capital on investments that go into the rate base if they incur large losses on other regulated activities, if they are to earn the cost of capital on average.

Pipeline Risks

We have concluded that gas pipelines are definitely *not* a low-risk industry. At the same time, it is not possible to characterize readily the level of risk currently facing the industry in any simple, dispositive way. There are two problems: first, the risks of merchant service probably differ from (and exceed, at present) the risks of pure transportation service; second, there is an implicit need for a baseline.

We have concluded that the risks of pure pipeline transportation service are up relative to any baseline. The two sources of this greater risk are increased competition and tighter regulation.

The FERC is apparently no longer inclined to permit the certification process to protect existing pipelines from new entry and new pipelines from mistaken entry. This raises the possibility of stranded investment, in which no set of rates will recover the approved cost of service. The experience of oil pipelines, where unfettered entry has been the rule for decades, confirms that new investments can fail and old investments can be rendered unprofitable by new entry despite the physical scale economies that seem to make pipeline transportation a natural monopoly. While oil pipelines are exposed to more forms of competition than gas pipelines, that is a difference in degree, not in kind. Thus, open competition for new transportation investments can be expected to expose gas pipelines to the risk of substantial losses, too.

At the same time, the regulation of transportation seems to be growing tighter. Throughput targets are being set more stringently than formerly, for example. The combination of competition and tight regulation *inevitably* leads to asymmetric losses. Thus, we conclude that transportation risks are definitely higher than they have been before.

The problem of a baseline becomes important in the assessment of merchant service and the overall risk of merchant and transportation service together. We have no doubt that the pipeline industry faces materially greater risks today than it did in the pre-NGPA days when the biggest problem was to find gas to satisfy the assured demand created by wellhead price regulation. The gas industry as a whole, from wellhead to burnertip, is riskier than it once was, and changes in the regulatory system have placed more of that risk on pipelines' shoulders.

If the baseline is the 1980s, however, those who argue for low risk are correct, *absent comparability transition costs*, that some of the important merchant-service risks of that decade are unlikely to surface again to the same degree.[12] New gas contracts, to the extent they exist, will be written with the experience of the past in mind. Also, Gas Inventory Charges, if implemented properly, could *in principle* compensate for the risks that remain, although practice has a long way to go to implement this principle.

Nonetheless, a transition to 'comparability' in firm sales under the mega-NOPR raises a new set of issues. The proposed rule is only in draft, and will undoubtedly be revised before final issue. Many details remain to be worked out. Thus, it cannot be analyzed definitively here. Still, there are four observations we can make.

1. Some pipelines may thrive in competitive gas sales markets, particularly those with little or no transition costs. Others, caught with long-term contracts preserved for a suddenly reduced merchant function, are at risk of a replay of the 1980s. On average, however, pipelines are likely to lose under comparability, because they already hold the remaining firm sales market and the first targets of competitors will be the most profitable parts of that business.

---

[12] Of course, the recent Columbia Gas bankruptcy makes clear that portions of the pipeline industry are still feeling the effects of the first round of restructuring, even as comparability begins.

2.   The ability to recover the costs of further con-
        tract restructuring is a major risk from compara-
        bility. It could also strand some investment in
        gathering lines. While the mega-NOPR suggests
        the Commission *intends* to allow pipelines to
        recover a substantial fraction of such costs,
        competition in the post-comparability market-
        place could easily render this promise empty.
        Moreover, since pipelines have not received
        compensation for the risk they might have to
        bear comparability transition costs, *any* share of
        the transition costs is too large a share.[13]

3.   Some features of the mega-NOPR may be attrac-
        tive to pipelines. For example, the proposal to
        shift to a straight fixed-variable rate design *may*
        reduce pipeline risks.[14] However, this is of
        value to pipelines only if the allowed rate of
        return on equity is *not* reduced corresponding-
        ly.[15]

4.   A host of regulatory and operational details
        remain to be worked out. Will profits or losses
        on pipelines' sales business be set off against
        allowed returns in transportation? How will

---

[13]   Pipelines make no profit on the gas itself. Gas Inventory Charges (GICs) were
        intended to compensate for the risks of signing contracts to meet firm sales commit-
        ments, but GICs at present are ill-defined and not widely implemented, and existing
        GICs in any case were not designed to compensate for the transition risks that arise
        under comparability.

[14]   Currently, most pipelines have to recover some of their fixed costs from commodity
        rather than in annual demand charges. The mega-NOPR apparently would shift
        more of the fixed costs to demand charges, although the details are not clear.

[15]   FERC staff have proposed reduced rates of return on equity if straight fixed-variable
        rate structures are used. See, for example, the testimony of Franklin D. Knight
        (1991). Historically, changes in rate design have been a response to broader chang-
        es. A switch to fixed-variable rates implies less risk only if all else remains equal,
        and that is not usually the case.

storage and other unbundled pipeline services be
priced and/or allocated among the pipeline and
other users -- users which in most cases will be
the pipelines' competitors in the end-of-pipe
market for natural gas? How can the system
best be operated under the new rule, and who is
at risk for the costs and inefficiencies that will
inevitably arise until this is worked out? The
NOPR to implement full comparability opens up
dozens of rules and procedures for argument and
maneuver by all the interested parties. Standing
in 1992 there is no way of saying which constituencies FERC will bless, which itself is a material risk for the industry today.

The balance of increased transportation risk and uncertain, but possibly
equivalent, merchant risk leaves the exact level of pipeline risk perhaps
roughly comparable to the 1980s, and possibly even higher. We have no
doubt that the industry today remains exposed to substantial risks overall.
The most important include uncertain regulatory rules and growing competition (including firm sales comparability), as well as 'wild cards' such
as environmental cleanup liabilities.

Regulated industries generally learned during the 1980s that regulation
does not protect against massive losses, and pipelines are certainly no
exception to that discovery. Traditional regulatory procedures do *not*
offer fair compensation for the risk of such losses. Nor do claims that
explicit changes in regulatory policy and the nature of the pipeline business now make pipeline risk low withstand scrutiny. That those changes
might have done so if everything else had remained equal is irrelevant,
because everything else did *not* remain equal.

A possible response to such problems is a loosening of the regulatory
constraint, to give pipelines some upside potential to balance the downside risk. Yet the picture we perceive is one of *tighter* regulatory constraint, through the use of more stringent throughput forecasts to set
rates, for example. Just as importantly, rule changes have been sufficiently frequent that pipelines must fear that even 'safe' investments may

become 'risky' on short notice.[16] The mega-NOPR clearly has this potential, for example.

Growing competition is another major source of risk. Importantly, proposals now exist to increase competition still further for both the transportation and sales parts of pipelines' business. Even if these proposals are not implemented, pressure for increased competition from both regulators and the marketplace is likely to persist. Competition ordinarily creates the potential for large losses; for example, a competitor may usurp a company's position with long-time customers. In principle, if competition is the only constraint, it provides an offsetting opportunity for large gains -- the business one pipeline loses can be another pipeline's gain.

The major problem that competition creates for pipelines, however, is that competition is *not* their only constraint. Pipelines increasingly face two masters, the market and rate regulation. Most of American industry faces one or the other, but not both. These masters discipline rates differently, creating the potential for large losses without the offsetting potential for large gains.

Thus, we find we side with the pipelines' view on their risks on balance. We have concluded that the pipeline industry cannot fairly be characterized as 'low-risk' overall, and in some dimensions -- particularly exposure to regulatory rule changes and to a mix of regulation and competition -- it is definitely high-risk. The view that the industry's troubles are in the past and that regulators can be counted on to fix any remaining problems is simply not borne out by the evidence.

---

[16]   For example, Order 451, made previously safe, cheap, 'old' gas risky. Producers got the right to begin renegotiations of gas contracts with pipelines, in the hope of raising prices, but pipelines could only put contracts on the negotiating table after the producer initiated the process. The result was that only producers with a larger share of old, cheap gas than new, expensive gas opened negotiations with a particular pipeline.

Long-Run Outlook for the Pipeline Industry

The most fundamental findings of our book, however, do not involve a ranking of the current level of risk facing the industry. More important questions are:

- Given the level of risk to which the industry is now exposed, whatever it is, do current regulatory and legal procedures permit investors to expect fair compensation for that risk? In other words, is the current system *sustainable*?

- Regardless of whether the current system is sustainable, are there aspects of the current system that should be changed? In other words, is it *desirable*?

Our answer to the first question before the mega-NOPR was a yes in the short run, providing material adverse rule changes were avoided, coupled with serious concern for the long run. The mega-NOPR has the potential to create serious new problems for some pipelines even in the relatively near term. However, the system is generally sustainable in the short run, absent rulings that put existing investment at risk, because pipelines have substantial sunk investment that will not vanish overnight.

In the long run, we find it highly unlikely that the level of new investment will be sufficient to maintain the national pipeline system at the level which regulatory and governmental policy makers may wish. The reason is our answer to the second question: there definitely are aspects of the current system that should be changed.

The first such aspect is that current regulatory procedures do not offer compensation sufficient to justify much risk-taking by pipeline management. Pipeline management now has the power to refuse to make investments where the potential reward does not justify the risk. If pipelines expect to earn only the cost of capital (the minimum necessary return to bring forth investment) if the venture succeeds but less than the cost of capital if the venture fails, they will only invest in ventures without material downside risk. For example, it is safer to build pipelines with less than economically optimal capacity and expand them when demand is certain than to size them optimally in the first place.

The problem is analogous to one noted by Judge Stephen F. Williams of the D.C. Circuit Court of Appeals in the context of contingent fees for lawsuits.[17] Judge Williams noted that the size of the markup courts permit over the noncontingent fee level determines the limit of the riskiness of the contingent fee lawsuits that are brought. A markup of zero would cause plaintiffs' attorneys only to sue if the suit were a guaranteed win. A markup of 100 percent would cause plaintiffs' attorneys to sue when the odds were 50-50, so that if ten cases were brought, attorneys would expect to earn twice the noncontingency fee in five of them, which is equivalent to the normal fee on average in each of the ten cases.[18]

As Judge Williams (1991) notes, the same forces govern behavior by regulated companies. If regulated investors expect only their cost of capital if the venture becomes used and useful in service to the public, but expect substantially less if something goes wrong, they will invest only where nothing major can go wrong. This is unlikely to be the optimal level of investment from society's viewpoint, because it implies pipelines will err persistently on the side of undercapacity.

The problems created by the current deterrents to risk-taking are exacerbated by the second problem that needs action: regulation that bases rates on an original cost rate base is incompatible with competition. As noted above, regulation overprices capital assets early in their lives and underprices them later on. Major new pipeline investments therefore cannot be undertaken in a competitive environment unless they can be 'rolled in' with old assets and unless the company offers some services where it can make up the competition-driven shortfalls from approved rates for other services. The result is guaranteed to be a socially suboptimal mix of pipeline investments, and almost certainly to be too little investment overall.

---

[17]   *King* v. *Palmer*, 906 F.2d 762, 769-70 (D.C. Cir. 1990) (concurring opinion). Judge Williams (1991, 159, 161) explicitly notes the analogy to the issue raised here in his review of a paper by two of the authors of this book.

[18]   This assumes, as Judge Williams does, that the risk of loss is entirely diversifiable or that the attorneys in question are risk-neutral.

These problems will not be fatal in the short run, because the industry's existing sunk investments will keep things running for now. That does not justify inaction, however. Systems that take a long time to decay will afterwards take a long time to fix, as students of the nation's railroads -- and its roads and bridges -- can attest. Moreover, there is a substantial economic cost in substandard or suboptimal service in the meantime. The authors of the National Energy Strategy have concluded that

> Despite . . . environmental advantages, ample supplies, and low wellhead prices, natural gas consumption has lagged. . . . [A] primary obstruction to increased utilization of gas is a regulatory morass that, through delays, distortions, and uncertainties, creates an atmosphere that is not conducive to the investment decisions necessary for producers, transporters, and consumers of natural gas to expand the markets. . . . Additionally, natural gas pipeline rate design policies *may deter construction of new pipelines and may not provide sufficient economic incentive to pipelines that are built to offer capacity at prices that would encourage maximum efficient use of natural gas.* . . . While wellhead supplies of natural gas are plentiful, *there is not enough pipeline capacity to satisfy demand in a number of regions* (U.S. Department of Energy 1991, 86-87, 91, 93, emphasis added).

Our findings reinforce these concerns.[19] If pipelines are to make economically sensible investments, they must believe their investors will be treated fairly in the long run. Fair treatment of investors and economically sensible pipeline investments *cannot* co-exist under a combination of competition plus tight regulation with an allowed rate of return equal to the cost of capital. Under such a system, pipeline management eventually *must* either (1) make investments that benefit customers but penalize shareholders, or (2) protect shareholders by refusing to invest. Maintenance of such a regulatory system is not sound public policy.

---

[19] The popular press contains frequent references to the lack of demand for natural gas despite depressed prices (*Wall Street Journal* 1991b, A1).

THE CHANGING PIPELINE INDUSTRY: FINANCIAL TRENDS

Appendix B sketches the recent history of the interstate gas pipeline industry more fully. That history reveals an industry undergoing substantial change. In the remainder of this chapter, we flesh out that picture with data on the industry in recent years. Most of this information comes from INGAA data bases, in part from a survey of pipeline companies that INGAA undertook for this study.[20] The remainder of the data comes from public sources.

For organizational purposes, we divide the data into two categories, 'financial' and 'other.' Data in both categories can have implications for several different types of risks faced by pipelines. Thus, a more detailed division did not seem useful. Our purpose in this chapter is to present information, not to analyze it. The analysis will come in Chapter 8, after Chapter 7 describes competing views of the industry's current status. Nonetheless, we do provide some brief comments to put the information in perspective.

The dominant financial trend in the pipeline industry in recent years has been the losses from the take-or-pay problem. As a result of Order 380 and subsequent decisions by the FERC and the courts, pipelines have had to buy out or otherwise renegotiate many contracts for high-cost gas.[21] Of the $8.8 billion in payments to producers which had been made as of September 1989, pipelines had absorbed $3.4 billion themselves.[22] Although exact data are difficult to obtain, nineteen pipelines were still exposed to an estimated $2.3 billion in take-or-pay liabilities as of September 30, 1989.[23]

---

[20]   INGAA conducted the surveys throughout 1990 and updated where possible in 1991. Most of the data in this chapter, therefore, are current through 1989 or 1990.

[21]   See Appendix B for a summary of the orders and events that led to these losses.

[22]   Order No. 500-H, Final Rule, FERC Stats. & Regs. ¶30,867 (Dec. 13, 1989), p. 45.

[23]   *Ibid.*

Several measures of the pipeline industry's financial trends confirm the difficulties the industry has been through in recent years. Figure 6-3 plots allowed[24] and actual rates of return on equity for several major interstate gas pipelines[25] from 1982 to 1990, the latest year for which data could be obtained. Strikingly, allowed rates of return on equity hardly moved at all during the years the pipelines were experiencing large take-or-pay losses. We will return to this phenomenon in Chapter 8, where we show that allowed rates of return do *not* track changes in the cost of capital closely.

Figure 6-3
Allowed and Actual Rates of Return on Equity

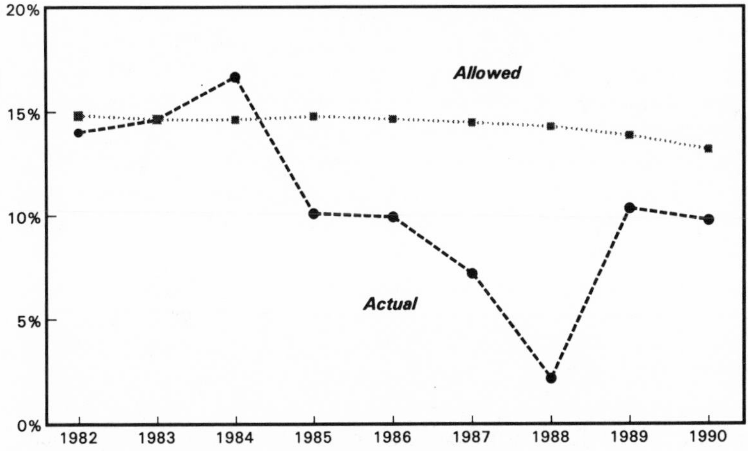

Sources: Allowed ROE: INGAA Risk Survey/Actual ROE: Net Utility OperatingIncome Survey.

---

[24] The allowed rates of return reflect all the allowed rates of return in effect for any given year. That is, if a pipeline receives a 13 percent allowed rate of return, that 13 percent is included in each subsequent year until a new rate of return is allowed in the next rate case.

[25] The collective market share represented by the sample of pipelines responding to the survey varied in each year.

Figure 6-4 compares the rates of return on equity that pipelines have earned to those of local gas distribution companies.[26] Information was available through 1990. For the period of comparison, however, pipeline earnings have clearly been more volatile than those of the other rate-regulated components of the gas industry.

Figure 6-4
Pipeline vs. LDC Actual Rates of Return

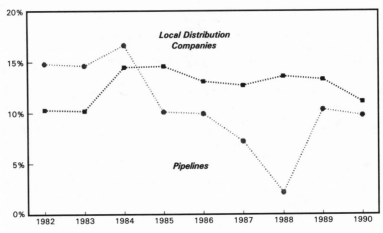

Sources: INGAA Risk Survey, Value Line

Figure 6-5 shifts to another measure, stock returns.[27] A problem in assessing the meaning of stock returns is that available stocks represent ownership of more than just gas pipeline companies. What we really want is a sample of gas pipeline 'pure plays,' a Wall Street term for stocks of companies in a single line of business. Four companies have 75 percent or more of their revenues derived from gas transmission in

---

[26] The figure does *not* imply LDCs were earning too much, but rather that pipelines were earning too little. See Figure 6-3.

[27] Stock returns are the sum of dividends plus capital gains divided by the beginning-of-year price.

1990: Panhandle, Sonat, Transco, and Williams.[28]   As a somewhat
broader measure, we also look at a sample that includes six additional
stocks, which have pipeline revenues of over 25 percent and but less than
75 percent of total revenues.[29]

Figure 6-5
Total Return for Natural Gas Pipelines vs. S&P 500

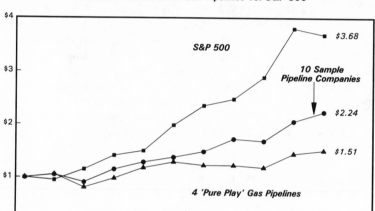

Sources: Value Line, Ibbotson Associates, Inc.   Stock Prices are Indexed to 1980 = $1

Figure 6-5 shows that pipeline stocks materially underperformed other
stocks in the bull market of the 1980s.[30]   On a compound annual basis,

---

[28]   Standard & Poor's Compustat Services, Inc., supplemented by Value Line 10/15/90
and discussion with KN Energy.

[29]   The six additional companies are Coastal, KN Energy, Columbia, Texas Eastern,
Consolidated Natural Gas, and Arkla. Data for 1989 pipeline revenues as a percent-
age of total revenues were obtained from Compustat, Value Line, and the Energy
Information Administration.

[30]   Of course, if pipeline stocks are of lower risk, we would expect them to do less well
than the general market in a period of market increases. We return to this question
in Chapter 8. (It turns out that pipelines are of *higher* risk.)

the Standard & Poor's 500 index appreciated at a rate of 13.9 percent
from 1980 to 1990, while gas pipeline stocks grew only at an 8.4 percent rate for the broad sample and a 4.2 percent rate for the four near
pure plays. Particularly noticeable is the wedge that opened in the 1985-
1988 period, when the magnitude of the take-or-pay problem hit home.

Pipeline bondholders as well as shareholders were affected by the events
of the last decade. Figure 6-6 plots pipeline bond ratings from 1980 to
1990.[31] Bond ratings provide a summary measure of the security the
rating agency[32] attaches to a company's debt. A drop in bond rating
(from A to BBB, for example) implies the timing and level of *future*
interest and principal redemption payments is not as secure as before.
The drop in bond rating may be due to a specific business risk or to
several factors. Since the claim of bondholders is senior to that of shareholders, a fall in bond ratings usually signals bad news for shareholders
as well. Lower bond ratings for a firm are associated with an increased
cost of raising capital. Figure 6-6 shows that the average rating dropped
in the mid-1980s and has not since recovered.

The figures to this point all addressed the year-by-year financial status
of the industry in one way or another. A somewhat different financial
measure is the rate base of the industry, which determines the size of
earnings under traditional rate regulation. (Recall Figure 6-1.) Figure
6-7 plots the industry's rate base in nominal and real (i.e., in current
and constant) dollars. In real terms, and largely in nominal terms as
well, the industry's rate base has declined since 1982.

It is impossible to correlate these numbers with actual pipeline condition
and capacity, however, since they are based on depreciated historical
costs rather than current net replacement costs. One rough indicator of
condition and capacity is the pipelines' reinvestment ratio. Figure 6-8
shows the percentages of pipeline annual cash flows that have been rein-

---

[31]   The average was calculated by assigning points to different bond ratings (C=1,
       CC=2, . . . AA+ = 17, AAA = 18) and by weighting the ratings among compa-
       nies in accordance with the size of their revenues. The companies are those among
       the sample of ten in Figure 6-5 for which information is available. Information was
       not available for each company in each year.

[32]   In this case, Standard & Poors.

Figure 6-6
S&P Bond Ratings

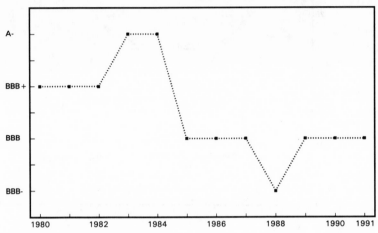

Source: Duff & Phelps Information on S&P Ratings

vested in the industry. This percentage dropped sharply in 1983 and
stayed low until 1990.[33] This has implications for the industry's financial condition, since a low reinvestment ratio implies a smaller rate base
and lower earnings, all else equal. It may also have implications for the
industry's ability to serve customer demand, either now or in the longer
run.

OTHER TRENDS

There are so many nonfinancial trends in the pipeline industry that it is
difficult to place them in order. The order we think will prove most
useful is as follows:

---

[33] We simply cannot say whether the higher reinvestment ratio in 1990 is an aberration
(e.g., a temporary 'catch up' for badly delayed investments in the lean years) or the
start of a new trend.

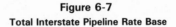

Figure 6-7
Total Interstate Pipeline Rate Base

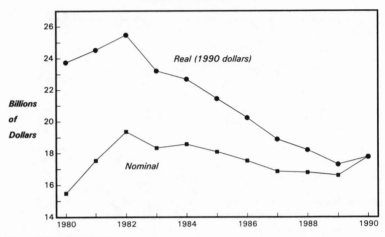

Source: T. Joyce Associates, Inc.

- The trend in overall gas demand.

- The shift from sales of pipeline-owned gas to carriage of gas owned by others.

- The resulting increase in operational imbalances between the amount of gas put into the pipeline and the amount taken out of the pipeline.

- The rapid swings between sales and transportation in the peak months.

- The lengthening of regulatory lag.

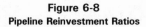

Figure 6-8
Pipeline Reinvestment Ratios

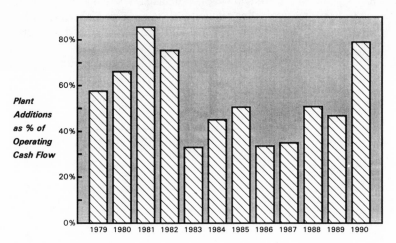

Source: T. Joyce Associates, Inc.

- The shift of fixed costs out of the 'commodity' charge and into the 'demand' charge.[34]

- The shift from firm sales and transportation to interruptible transportation.

- The growth of discounting.

---

[34] As discussed above and in Appendix B, pipelines in recent years have recovered some portion of their fixed costs in the charge for each unit of gas the customer buys, the commodity charge, and the rest in an annual fee that is independent of the amount the customer buys, the demand charge.

Gas Demand

Figure 6-9 plots the overall demand for gas transmission from 1972 to 1990.[35] The trend is up, showing that demand is stronger now than in the early 1980s when pipeline reinvestment ratios were higher. Over the next twenty years, demand is expected to return to the levels of the early 1970s.

Figure 6-9
Historic and Projected Demand for Natural Gas

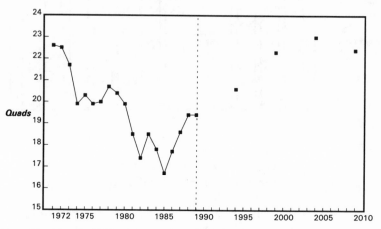

Source: Energy Information Administration, historical data: Monthly Energy Review (August 1991),
        projections: Annual Energy Outlook (March 1991).

Growth of Carriage

One of the most striking trends in the pipeline industry is the decline in the pipelines' merchant function, i.e., in the proportion of flowing gas to which pipelines take title. Figure 6-10 shows that in 1983, pipelines

---

[35] A 'quad' is a quadrillion Btu. A common rule of thumb is that there are about 1,000 Btu in one cubic foot of gas, so one quad of gas is roughly equal to one trillion cubic feet.

owned about 95 percent of the gas they transported. By 1990, they owned about 20 percent. Third-party marketers (including producers) owned more than pipelines, approximately 40 percent of the gas transported. The rest was owned directly by the pipelines' final customers, i.e., distribution companies and end users. In short, competition for the pipelines' traditional sales function is now widespread.

Figure 6-10
Merchant Function Providers

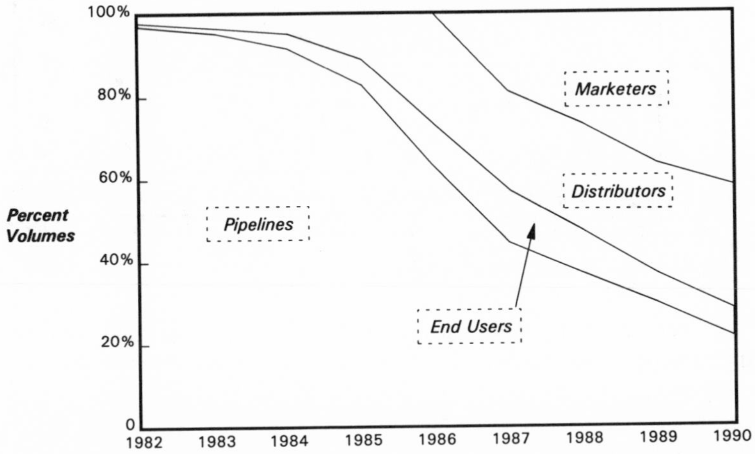

Source: INGAA Report, 'Carriage Through 1990'

Figure 6-11 shows the trend in volumetric terms rather than percent. From 1982 to 1990, carriage grew from roughly half a quad to about 13 quads, while sales declined from over 13 quads to about 4 quads. Thus, most of the growth in demand discussed above has been satisfied by carriage, not sales.

Figure 6-12 shows the number of carriage transactions (i.e., individual scheduling of gas movements). They increased more than fourfold from 1985 to 1988 and have remained roughly constant since then. Before open-access transportation, pipelines controlled access to the system and had to deal only with a handful of distributors and end users. Now they

Figure 6-11
Sales vs. Carriage for Market

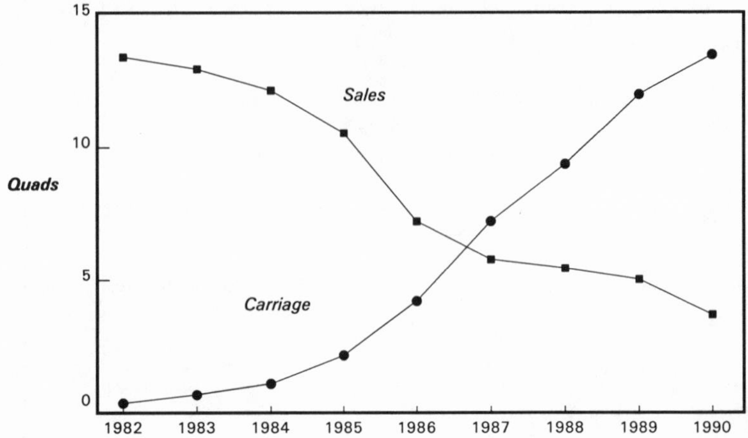

Source: INGAA Report, 'Carriage Through 1990'

must deal with many more entities, each trying to put gas into the pipeline or take it out.

Imbalances

Imbalances between the amounts of gas put into and taken out of the
pipeline flow directly from the rapid growth in the number of transactions, and are increasingly of concern to the pipeline industry. Existing
equipment generally does not permit real-time monitoring of the gross
amounts put in at each collection point nor the amounts taken out at each
distribution point.[36] Moreover, while the aggregate amounts are eventually known at each point, if there is more than one supplier at a collection point or more than one customer at a distribution point, there may
be no way to be sure who has not performed as planned. Finally, if the

---

[36]   The mega-NOPR on comparability may require installation of such equipment, but
it is not generally in place at present.

Figure 6-12
Count of Carriage Transactions at Year-End

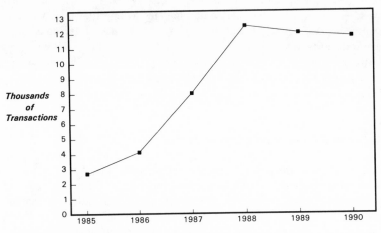

Source: INGAA Carriage Reports, 1986-1990.

pipeline is not a party to the supply contract, the pipeline may not even know who are the candidates for nonperformance.

The December 1989 cold front provides an example of the problems which may arise from imbalances and the increasing number of transactions. The December 1989 cold front left pipelines in a position of being owed more than twice the amount of gas than they owed others. At the end of December, five pipelines responding to an INGAA survey[37] reported that they owed 2.5 Bcf to others, while another five pipelines reported they were owed 5.6 Bcf by various parties. Each of the remaining twelve pipelines either did not have an imbalance, or did not have information available at the time of the survey.

The imbalances were created when pipelines were confronted by nonreceipt of nominated volumes at the supply end, and by gas taken without

---

[37] Twenty-two pipelines representing approximately 82 percent of the market share responded (INGAA 1990). The survey was conducted February-April 1990.

authorization at the demand end. At the supply end, 44 billion cubic feet (Bcf), amounting to 16 percent of nominated volumes, were not received by pipelines from December 23 through December 27, the most severe stretch of the cold front. At the market end, 6.7 Bcf was taken from pipelines without authorization during this period. These and other weather-related problems led some pipelines to interrupt transportation volumes and ultimately to curtail some sales volumes. Interrupted sales and transportation volumes totaled 48 Bcf for December. Sales volumes, however, were relatively well protected, as pipelines curtailed an average of less than one percent of the total reported contract demand for December.

Concentration of Sales in Peak Days

Another trend is the concentration of the pipelines' merchant function in the peak days. Figure 6-13 shows the proportions of actual annual and peak sales as a percentage of the pipelines' service obligations for annual and peak sales. The proportion of annual sales actually made in 1989 (a bit over 30 percent) was less than half the proportion made in 1980 (a bit over 70 percent). Yet the proportion of peak sales had declined hardly at all (from just over 100 percent to a bit under 90 percent).

The service obligation requires that pipelines balance their supply portfolio with their demand portfolio. Since annual takes have declined as a fraction of annual obligations while peak takes remain a high fraction of peak obligations, pipelines must keep far more gas 'at the ready' than the current competitive situation calls for. Moreover, pipeline obligations to serve are open-ended and extend beyond the end of the current contracts. Thus pipelines cannot just wait for this problem to go away.[38]

Of course, overall sales have been declining with the growth of carriage. But the pattern remains even when looking at sales and carriage individu-

---

[38]   Gas Inventory Charges (GICs) are intended at least in part to deal with such problems. However, GICs are taking a long time to implement, and there is no guarantee that they will be compensatory when implementation finally comes. See Appendix B and Chapter 8. Comparability ultimately may relieve pipelines of their service obligations, but the timing and even the event of such relief is uncertain at present.

Figure 6-13
Relationship of Actual Takes to Service Obligations

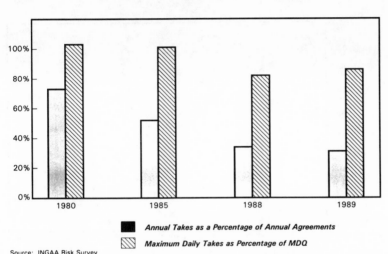

Annual Takes as a Percentage of Annual Agreements

Maximum Daily Takes as Percentage of MDQ

Source: INGAA Risk Survey

ally. Figures 6-14 and 6-15 show this in slightly different ways. Figure 6-14 shows the percentage of sales to total volumes during peak and off-peak periods from 1984 to 1990. Sales have been a declining proportion of total volumes both peak and off-peak, but the decline has been much greater off-peak.

Figure 6-15 plots the quarter-by-quarter proportion of carriage over the same period. Carriage is clearly concentrated in the warm, off-peak quarters (the second and third) while sales are concentrated in the cold, peak quarters (the first and fourth).

In fact, even within the peak month, sales are increasingly concentrated on the peak days. Figure 6-16 shows a growing ratio of maximum daily sales to average daily sales within the peak month. This indicates that swings to and from sales have increased not only seasonally, but within a heating season. That is, customers now swing back and forth more in the peak month than they did before. Rapid swinging is difficult operationally. Risk is increased because the pipeline does not know when the

Figure 6-14
Seasonality of Sales

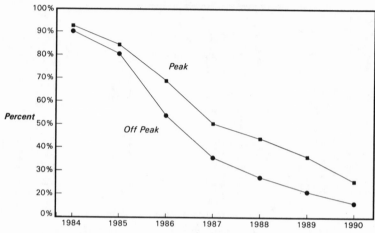

Source: INGAA, 'Carriage Through 1990'

swing will happen and can easily have either too much or too little of its own gas in the system to match the swing.

Regulatory Lag

Two regulatory trends require mention as well. The first is the lengthening of regulatory lag. Figure 6-17 depicts the average number of months from the initial rate filing to a final Commission order before Order 380, in the first filing after Order 436, and in the most recent filing. It now takes almost 30 months to complete a case, nearly twice as long as it did before Order 380.

This creates uncertainty for pipelines and their customers alike. When rates are proposed, they go into effect subject to refund. The longer the rate case drags on, the longer the uncertainty over how much customers actually will pay and how much pipelines actually will have earned.

Figure 6-15

Carriage Share of Total Delivered for Market Through 4th Quarter, 1990

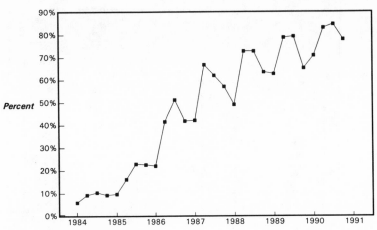

Source: INGAA Report, `Carriage through 1990'

Shifts of Fixed Costs

The second area of regulatory change is the way the pipeline's fixed (i.e., not volume-sensitive) and variable (i.e., volume-sensitive) costs are recovered through demand and commodity charges.[39] Figure 6-18 depicts the proportion of fixed costs that have been included in the commodity charge under the major precedents. The first standard was set in 1938, when no fixed costs were in the commodity charge. The 'Seaboard' formula of 1952 split fixed costs 50-50 between demand and commodity charges. The 'United' formula of 1973 upped the proportion of fixed costs in the commodity charge to 75 percent, largely as a way to discourage consumption by raising prices in response to the growing shortfall of gas in interstate commerce resulting from wellhead price controls.

---

[39] See above for additional discussion of these terms and Appendix B for citations to the relevant FERC decisions.

Figure 6-16
Ratio of Max Daily Takes to Average Daily Takes of Peak Month

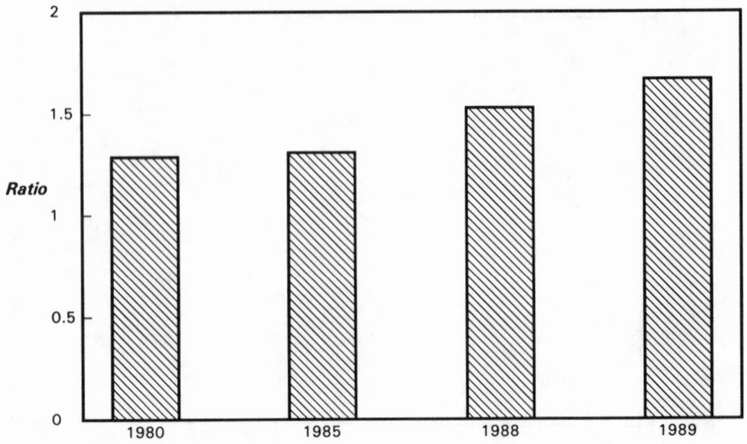

Source: INGAA Risk Survey

The 'Modified Fixed/Variable' (MFV) formula of 1983 was a reaction to the inability to market high-priced gas. It put all fixed costs except (1) return on equity and the associated taxes and (2) the fixed portions of production and gathering costs, in the demand charge. The proportion of fixed costs that remained in the commodity charge under MFV thus will vary from pipeline to pipeline, leading to the band of uncertainty shown in the figure.[40] However, MFV also splits the demand charge into two components, one based on peak usage and one based on total annual usage.[41] The annual usage component of the demand charge (i.e., D2) thus is variable with volume over a somewhat longer run as

---

[40]   The usual rule of thumb, according to former FERC Commissioner George R. Hall, is that 10 percent of the fixed costs remain in the commodity charge under MFV.

[41]   These are known as 'D1' and 'D2,' respectively. See Appendix B for discussion and Figure 6-2 for an illustration of how the D1 and D2 charges affect rates.

Figure 6-17
Average Number of Months From Initial Rate Filing to Commission Order

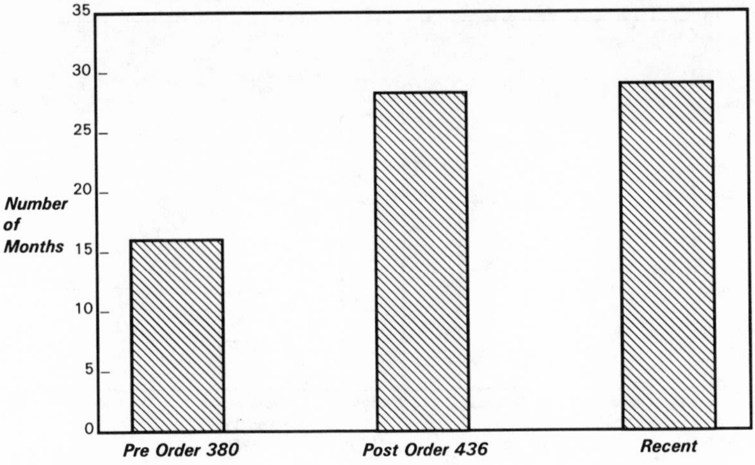

Source: INGAA Risk Survey

well. The issue of rate structure is currently once again under review by the FERC, so the future split is uncertain.[42]

Growth in Interruptible Transportation

Figure 6-19 breaks the 1990 carriage down by whether it is firm or interruptible service. It shows that 36 percent of the 1990 carriage was for firm service, while 64 percent was interruptible service. Firm service costs more and is bought to assure service during periods of peak demand.

---

[42] The mega-NOPR proposes to return to a straight fixed-variable rate design. Whether such a change is valuable to pipelines depends on whether an offsetting adjustment is made to the allowed rate of return. In practice, changes in rate design are often a response to other changes. Whether any adjustment to the allowed rate of return is warranted depends on how these other changes affect the pipeline's risk.

Figure 6-18
Proportions of Fixed Costs in Commodity Charge

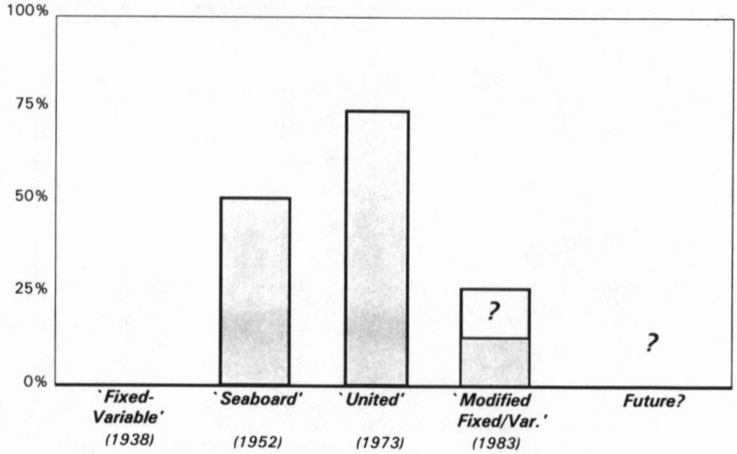

Source: Discussion in Appendix B

This high level of interruptible service represents a change, too; that is, not only the level but also the composition of carriage has changed. Figure 6-20 shows the approved and actual volumes of sales, firm transportation, and interruptible transportation in the rate case immediately before Order 380, the first case after Order 436, and the most recent rate case for the pipelines responding to the survey. Several points should be noted.

First, sales have declined and carriage (the sum of firm and interruptible transportation) has grown, as discussed above. Second, the level of sales has consistently been overestimated in the ratesetting process and the level of interruptible transportation has consistently been underestimated. Third, the amount of interruptible transportation has been growing dramatically. Finally, total throughput has consistently been overestimated when setting rates.

To interpret these trends, two others must be noted. The first is an important aspect of the MFV rate design and the volume trends in the industry: a growing proportion of fixed cost recovery depends on inter-

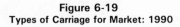

Figure 6-19
Types of Carriage for Market: 1990

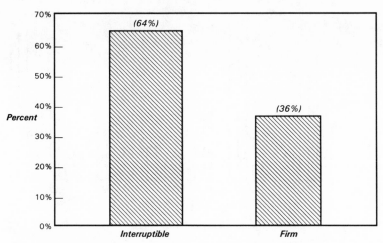

Source: INGAA Report, 'Carriage Through 1990'

ruptible service. Figure 6-21 shows the percent of fixed costs allocated to interruptible transportation in the rate case immediately before Order 380, the first case after Order 436, and the most recent rate case for the pipelines responding to the survey. Nearly 30 percent of pipelines' fixed costs now must be recovered from interruptible transportation. Moreover, since these proportions are based on *allocated* rather than *actual* volumes, the pipelines' actual dependence on interruptible transportation for fixed cost recovery is even greater. (Recall Figure 6-20.)

Growth of Discounting

The final trend is the growing importance of discounting. Figure 6-22 shows that discounting affected about 2.5 quads of gas in 1989, compared to virtually none in 1985. Although the figures are not strictly

Figure 6-20
Approved vs. Actual Volumes

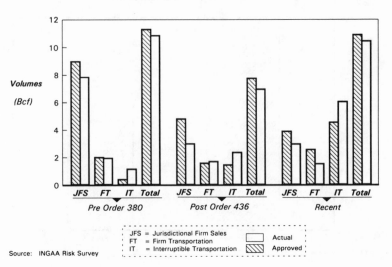

Source: INGAA Risk Survey

comparable, this is nearly half of the actual recent interruptible sales shown in Figure 6-20.[43]

Discounting is of concern to pipelines because it implies less revenue than if discounting were not necessary. Of course, if discounting did not enable pipelines to retain customers and thereby increase revenues in actuality, presumably no discounting would take place. Thus discounting is a response to competition of one sort or another, and the growth in discounting is evidence of a growth in competition.

Discounting is also important in assessing the ability of pipelines to recover their fixed costs and in the interpretation of Figure 6-20. Figure 6-20 might seem to suggest that pipelines are over-recovering their fixed costs, since the firm components of demand which include fixed demand

---

[43]   Figure 6-20 refers to the most recent rate case, not necessarily to 1989. Also, Figure 6-20 is in billions of cubic feet, while Figure 6-22 is in trillions of Btu. As noted earlier, it is common practice to treat these measures as equivalent.

Figure 6-21

Percent of Fixed Costs Allocated to Interruptible Transmission

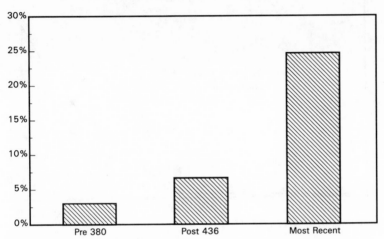

Source: INGAA Risk Survey (data for 5 pipelines).

charges, are overestimated when rates are set and interruptible demand is underestimated. However, given the increasing reliance on interruptible service for fixed cost recovery, the rapid growth of discounting may imply that fixed costs will not in fact be recovered.[44]

---

[44] At one point, the FERC did not recognize discounted volumes when allocating fixed costs to various classes of service, which virtually guaranteed under-recovery of fixed costs. They subsequently were ordered to set aside discounted volumes when calculating the maximum rate for interruptible transmission (*Interstate Natural Gas Pipeline Rate Design, et al.*, 47 FERC ¶61,295, *order on reh'g* 48 FERC ¶61,122 (1989), *appeal docketed, Wisconsin Public Service Commission v. FERC*, No. 89-1598 (D.C. Cir. September 25, 1989). This policy still does not apply to other services such as storage, which, because of competition, also require discounts. For such services, discounts lead to under-recovery. The result is a combination of greater-than-expected interruptible volumes, rapid growth of discounting, and uncertain rate treatment, which makes it difficult to decide as a matter of principle whether pipelines can expect to recover their fixed costs on average. As a matter of practice, however, Figure 6-3 suggests they have not done so in the recent past.

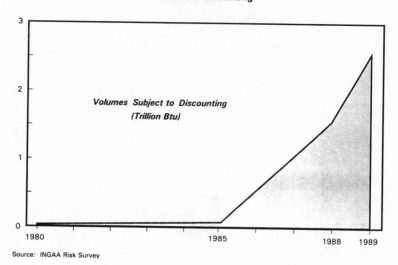

Figure 6-22
Growth in Discounting

Source: INGAA Risk Survey

CONCLUSION

The pipeline industry thus has been changing rapidly. It went through
hard financial times in the 1980s, and it remains subject to considerable
uncertainty about its future. Its traditional mode of operation has van-
ished, and new modes are evolving. Those modes involve fewer and
'peakier' sales and more transportation, particularly interruptible transportation, a more complicated operational environment, and more competition.

Two polar views exist on the present outlook for the industry. One is
that the industry is still exposed to substantial risks and that its future is
questionable. The other is that the industry's bad times are behind it and
it currently faces relatively little risk. The next chapter presents these
views in more detail.

Chapter 7
Two Views of Pipeline Risk

Since capital markets are self-correcting, the costs of capital for a utility operating in such an environment ['used and useful'] will be greater and the company will recover these costs through a higher rate of return.

. . . capital markets can be relied upon to identify the risks faced by utility investors and to compensate them *ex ante* for taking them. It is widely accepted today that when investors purchase publicly traded securities in a company they do so at a price which can be presumed to reflect all publicly available information. That price thus implies an expected return which is adequate by definition, to compensate investors for the known risks which the company faces. Can these risks be identified? . . . These risks should be comparatively easy to identify.

*　　*　　*

When that rule ['used and useful'] is incorporated within a particular regulator's "methodology," it shifts the risk of unsuccessful, albeit prudent, utility investments to investors, who are automatically compensated for taking this additional risk by the operation of capital markets (Goldsmith 1989, 266, 272-273, footnotes omitted).

INTRODUCTION

In Chapter 6, we sketched the recent history of the interstate gas pipeline industry. That history is developed more fully in Appen-

dix B. We doubt that anyone reading that history would disagree
with the conclusion that the natural gas industry, and particularly
its pipeline segment, has experienced a much higher degree of
risk since the Natural Gas Policy Act passed in 1978.   Legal
and regulatory standards and contractual and business relationships
have been in flux.   Pipelines have had losses of billions of
dollars from take-or-pay contract litigation and settlements.   Competition for what once were protected markets not only is growing, but is being actively encouraged by the regulatory agency in
control, yet regulatory jurisdiction over rates remains in force.

Even over a longer horizon, the natural gas industry has evi-
denced great uncertainty.   The system has experienced major
upheavals from time to time, starting with the *Phillips* decision
to regulate wellhead prices.   Energy prices in general have been
subject to major swings as markets accommodate world events,
and those swings have in turn driven many of the legal and
regulatory actions, as Congress, regulators or the courts have
stepped in to try to fix the problem once and for all in an
acceptable way.   In each case to date, the 'fix' has set the stage
for another round of problems.

The regulatory scheme, in particular, in recent years has been
characterized by uncertainty and instability.   Deliberations in major
rule makings are lengthy and even 'final' agency decisions are
rarely indeed final.   Major policy changes are routinely remanded
by the courts for further lengthy decision making and these orders
are often themselves remanded yet again.   Order 451, which
effectively decontrolled the price of 'old' gas, took nearly five
years to clear the courts and did so only after the Supreme
Court reversed a lower court's ruling.   The Order 436/500 process is likely to take a decade to develop a clear set of ground
rules for the new regulatory and competitive regime.

Thus radical, unexpected changes have not been unusual. These changes have initiated great shifts in wealth among the participants along the vertical chain from wellhead to burnertip. The question today, for investors and regulators alike, must be: is the system finally sustainable, or is it just a matter of time before the next round of wealth-shifting change?

A person's view of the current level of risk facing the industry is likely to be colored heavily by his or her answer to this question. Thus, we also noted in Chapter 6 that while the industry perceives risks are high, the industry's regulators and customers appear to believe that pipeline risk is now rather low, either relative to other industries or relative to this industry's immediate past. Logically, a level of risk that is higher than it was pre-NGPA does not necessarily imply that pipeline risk is higher in some absolute sense than in other industries. Nor does a higher level of risk over a long period necessarily imply that risk today is as high as it was a few years ago.

The view that risks are low reflects in part the belief that the system is in fact now on the way to a sustainable equilibrium, a system under which the various segments of the industry can survive and perhaps even prosper. In particular, those who hold this view appear to feel the major risks are behind the industry and that changes in the industry itself and in the rules by which it is governed now point the way to a more secure, stable period.

Before turning to the details of these views, the focus of this chapter, it is useful to consider what issues the industry still faces. It will also be helpful to distinguish between the total risk facing the natural gas industry and the incidence of that risk.

Issues Outstanding

An analysis of pipeline risk can start with a litany of unresolved problems in the transition toward a more competitive gas pipeline industry.[1]   While FERC orders may address the basics of these issues, many are not yet settled, despite the passage of at least six years and billions of dollars at stake.

The Legacy of Federal Wellhead Price Controls.   Although the FERC attempted to resolve the issue of deregulation of 'old gas' in Order Nos. 451/490, court review was until recently still up in the air.   The Supreme Court ruling upholding Order 451 may now have settled this issue.   The result confirms that the prices on old gas contracts can be raised, but without giving pipelines an equally easy mechanism to lower prices on newer, high costs contracts.   This discrepancy has worsened the take-or-pay problem since Order 451 was first promulgated.   Full deregulation in 1993 should resolve the price issue for future contracts, *providing* legislatures and regulators convincingly forswear any intention to intervene once again.

The Legacy of the Take-or-Pay Contracting Problem.   Much of this problem has been solved, but many old contracts are still being implemented, and many contractual disputes are still being litigated and negotiated.[2]   The ability to recover everything but

---

[1]   We are indebted for our starting point to Jensen Associates, Inc. (1990), who raised the first five of these issues.

[2]   The recent Columbia Gas bankruptcy demonstrates that the problem is not fully resolved.   The company's July 31, 1991 press release stated that

the pipelines' 'share' of the problem through the Order 500 process remains an uncertainty as well.

The Conflict between Cost-of-Service Regulation and Market Pricing. If customers have the right to a 'just and reasonable' price while retaining the ability to 'play the market,' the regulated price becomes a ceiling price. The result may turn out to be a downward-biased ratemaking approach. Unless the pipeline is allowed to depart upward from cost-of-service regulation to make up the difference, the result could inherently guarantee inadequate returns. Moreover, this problem is exacerbated by the intrinsic differences between regulated and competitive prices, a topic we discussed in Chapter 4.

Within the third area of uncertainty are a host of rate design issues that are as yet unclear in their implementation:

    a.   Scope of permissible discounting and treatment of revenue dilution from discounting in the revenue requirement;

    b.   Ability to impose Gas Inventory Charges (GICs) (and whether GICs will actually compensate for the risks pipelines bear; see Chapter 8 and Appendix B), or, under comparability, the ability to receive adequate compensation for contracting risks in other ways;

"above-market-priced gas contracts . . . are the root of [Columbia's] financial difficulties . . . ."

c.   Numerous rate design issues, such as mile-
age-based rates, unbundling of storage and
other services and costs, seasonal rates,
rolled-in versus incremental tariffs, etc.;

d.   Access to firm capacity (who gets it),
capacity brokering (can it be resold), and
capacity entitlement (rationing interruptible
transportation, etc.);

e.   The future form of pipeline and producer
revenue guarantees (contract and tariff mechanisms to shift risk); and

f.   The certificate mechanism for approving new
pipeline construction.

***How Well a System Initiated during Surplus Will Function under
Balanced or Tight Market Conditions.***   It is unsurprising that
pipeline customers today seem reasonably satisfied with a system
of low prices and bountiful supplies.   However, if gas markets
ultimately show the same volatility as oil markets, there may be
demands to repeat the entire regulatory cycle.

***The Conflict between Sanctity of Contracts and Protection of
Small Customers.***   Key to the current approach to deregulation of
gas markets is the requirement that parties to contracts be required to live with the outcome.   Since 'mistakes' (viewed retroactively) are inevitable in a high-risk industry, there will equally
inevitably be pressure on legislators and regulators to intervene to
protect favored constituencies from unfavorable outcomes.   Whether
policy makers will be able to stay the course when favored con-

stituencies turn out to be 'backing the wrong horse' remains to be seen. In the past, gas supply curtailments have led to rationing schemes to benefit priority users instead of to reliance on market mechanisms (i.e., rationing by price). A possible trouble spot today is the tendency of distributors to use spot markets heavily except during peak periods, while pipelines are reluctant to sign long-term gas supply contracts in the present regulatory environment.

Clouds on the Horizon. A number of issues are not definitely problems for the industry, but represent 'things to worry about.' For example, there is a bill before Congress to increase pipeline competition for the *transportation* function. Environmental risks are another concern. They range from cleanup of past wastes to possible regulations on methane loss (unburned methane is a greenhouse gas). Gas quality is another concern, since low-quality gas injected without a pipeline's knowledge drags down the average quality of all the gas.

Finally, comparable access requires mention. It has gone from a cloud on the horizon in early 1991 to a near certainty with the mega-NOPR in July 1991. *Comparable access* or *comparability* means unbundling pipeline services to such an extent that anybody can offer the same quality merchant function as the pipeline. That is, the pipeline would have no material advantage over any other offerer of the merchant function. The chief proponents of this concept are marketers, who would, as unregulated merchants, be able to charge what they wanted for the merchant function while the pipeline is stuck with high-cost contracts on the supply side. Essentially, marketers would like to have storage rights, the right to switch to another producer when the first producer fails to deliver, and so on, in short, the right to any service or facility the pipeline uses in fulfilling its merchant function.   This concept clearly increases competition for pipelines.

FERC seems amenable to allowing the 'lighter hand of regulation' to pipelines who accept this concept.   Unfortunately, comparability touches upon many major issues, such as pre-granted abandonment of the sales service obligation.   The marketers, including large producers, would undoubtedly like pipelines to maintain low-load factor customers, of course, because that would leave the pipelines with the riskiest customers.   The comparability concept is the heart of the mega-NOPR discussed in Chapter 6.

Surprises.   To the previous list we would add a final risk: 'unknown unknowns.'   Risk often comes as a surprise, from something no one thought to worry about, or at least from something that was thought to be such a remote possibility that it made no sense to worry about it.   For example, the energy price shocks of the 1970s came as a surprise to many, as did the energy price collapse of the 1980s.   The former contributed heavily to the need for the NGPA to overturn the effects of the *Phillips* decision, while the latter contributed heavily to the industry's take-or-pay problems.   Any number of important but unlikely 'surprises' might be imagined today, from practical cold fusion to a long-term disruption in Middle Eastern oil supplies.   While it is impossible to say if any of them will actually materialize, neither can the generic potential of unknown unknowns be ignored.

Total Risk Versus Risk Incidence

A key distinction in any discussion of the level of risks facing pipelines is the difference between the total risk facing the natural gas industry and the incidence of that risk on the various components of the industry. Regulatory rules and voluntary contracts affect the allocation of risk among gas pipelines, their suppliers, and their customers. Indeed, much of the history of recent regulatory experience in the gas pipeline industry, as sketched in Appendix B, could be characterized as an effort to reassign risk in light of events outside the control of regulators.

Recent literature (e.g., Legato 1987) on regulatory institutions has addressed the issue of which party is best suited to bear risk of regulated industries, both up and down the vertical chain of buyer/seller relationships and within and outside the industry. Examples would include the use of various contracting mechanisms for allocating risk among buyers and sellers (e.g., take-or-pay contracts) and regulatory mechanisms to accomplish the same result (e.g., fuel adjustment or purchased gas adjustment clauses). Our study does not attempt to apply this literature to determine the optimal distribution of risk among affected parties, but rather recognizes that the existing distribution is determined by the existing regulatory and contractual institutions.

The ability of contracts and regulatory mechanisms to allocate risk up and down the vertical chain from wellhead to burnertip is important. The allocation of risk to pipelines could be increasing or decreasing as a result of changes in these institutions regardless of whether or not total system risk (i.e., from wellhead to burnertip) were increasing or decreasing.

Compared to 'the good old days,' it is clear that total risk in the gas industry has increased enormously. The issues that govern the debate over the pipeline industry's risk then are: (1) how much of this increased risk is now borne by gas pipelines; (2) how much has been passed forward to pipeline customers (via changes in rate design and switches from tariff gas to transportation); (3) how much has been passed back upstream to gas producers via 'market out' contract provisions and purchases of gas on spot markets rather than long-term fixed-price contracts; and (4) how secure are the contractual and regulatory mechanisms that have effected this allocation of the gas industry's overall risk? Precise answers to these questions are not achievable, but the issues they raise are vital to an understanding of the impact of rule changes and economic events on pipelines' risks.

The Remainder of the Chapter

Since regulators control the rates of return pipelines are allowed and the rules under which pipelines operate, a misperception by regulators of the level of pipeline risk and the sustainability of the evolving regulatory system could have serious consequences. An excessive perception of risk would overcharge customers and unnecessarily discourage gas use, while an insufficient perception of pipeline risk would discourage the amount and kinds of investment that pipelines should make.   Even more fundamentally, even if the system proves sustainable, an important policy question is whether *this* sustainable system is *desirable*.

In this chapter, we focus on the first issue, whether pipeline risk is still high today, or whether instead pipeline risk can now be fairly characterized as 'low.'   In particular, this chapter presents our understanding of the differing views on that question.   Chapters 8 and 9 will present our own views on that issue and on the broader, and we think far more important, question of whether the system as it is evolving is desirable.

THE VIEW THAT RISK IS HIGH

The reason this study exists is because pipeline executives believe they face high risks, but perceive that others have different views. The natural place to begin the discussion of the view that risk is high is thus with the perceptions of the industry itself.

Types of Risks Perceived in the Business World

To start, we asked the Interstate Natural Gas Association of America (INGAA) to supply a list of risks (displayed in Table 7-1) as seen by the pipeline industry.

Table 7-1 PRELIMINARY INDUSTRY LIST OF RISKS	
Primary Risk	**Related Risk Area**
Service Obligation	Contract Demand Conversions/Adjustments
Cost Recovery	Take-or-pay Competition Contract Demand Conversions Return on Investment Recovery of Investment Recovery of Operating Costs Volume Setting Risk Equity
Cost of Capital	Credit Ratings Return on Debt Return on Equity
Operational Risk	Volume Reconciliation Lessened Control of System Product Quality
Market Risk	Discounting Bypass Loss of Market Share Alternate Fuels Marketer and Other Alternative Supplier Competition Optional Expedited Certificates
Regulatory Risk	Regulatory Delays Costs Recovery/Used and Useful Compliance Disclosure Changes in Policy/Commissioners

Source: Interstate Natural Gas Association of America, private communication.

When the average business person, in any industry, addresses the issue of risk, that person will usually ask "What can go wrong?" We think this description fairly characterizes the list in Table 7-1. The Primary Risks in the table are areas where trouble might start. The Related Risk Areas are specific factors that could make 'things go wrong.'[3] The ultimate impact of 'things going wrong' may be felt in projected rates of return, earnings levels, share prices, sales, or costs. 'Risk' is then viewed as potential changes in the assumptions about the future that would drive realized values of these variables far from their desired values.

For the pipeline industry, future 'bad events' that might adversely affect the company, from a business person's point of view, may be organized as follows:

1. *Regulatory risk:* Will changes in governmental rules, regulations or laws adversely affect revenues or costs?

    Example: Government regulations impose price controls that reduce revenues or impose environmental regulations that

---

[3]   Interestingly, comparability is not even mentioned in this list, which was supplied in mid-1990. The speed with which the Commission has moved toward comparability demonstrates how fast an 'unknown-unknown' risk can develop.

increase costs of transporting natural gas.

2. *Business risk:*   Will changes in markets limit sales or the prices that can be charged or increase the prices of inputs?

    Example: Energy prices decline, making it impossible to market gas at prices necessary to recover contract costs and earn a competitive rate of profit.

3. *Operational risk:*   Will the enterprise's operations perform as planned?

    Example: Severe cold weather causes equipment to freeze, causing curtailment of gas pipeline throughput.

Although some categories of risks may involve all elements of the classification, the distinction between these categories is clear in principle. Regulatory risk arises from actions of regulators or legislatures and the courts that review their actions. Business risk focuses on the firm's relationships with customers, competitors, and suppliers, while operational risk focuses on the mechanics of providing service.

Regulatory Risk. Recent examples where regulatory risk has proven to be a source of 'things that can go wrong' for pipelines are numerous. Order 380 paved the way for massive take-or-pay losses. Order 436/500 changed the entire nature of the pipeline business, from largely a merchant business to a mixed merchant and carriage business. The mega-NOPR may eliminate the merchant business entirely for some pipelines. The ability to build competing pipelines is being enhanced by FERC's interpretation of Section 311 of the NGPA. Appendix B describes this history in more detail.

Furthermore, actual practice may differ significantly from Commission statements of policy. There appears to be no enforcement mechanism to ensure that FERC staff inputs to the ratemaking process actually conform to the stated intents of regulators. Pipelines complain that their objections regarding discrepancies between practice and policy are answered by an observation that pipelines have legal redress for any inequities that result. This pattern often leaves one with a folklore of opaque practice, for which it is not easy to generalize the consequences for risk.

Indeed, we have found that attempting to assess the risks of regulatory policy on the gas pipeline industry is at times a frustrating experience. The end product of the regulatory process is often a 'black box' settlement with no ability to read or even to infer the principles embodied in the process. FERC policy on issues such as determination of the allowed rate of return are embedded in the numerous decisions and rationales, with no clear statement of a coherent methodology.

Business Risk.   Business risk is higher than it used to be, too. One reason is growing competition for both sales and carriage. Also, energy prices remain subject to sharp fluctuations based on world events.   Competition from a commodity with widely vary- ing prices leads to swings in demand that make both investment and pricing decisions difficult.

More generally, the sustained boom of the 1980s is over, and no one seems quite sure what the economic outlook for the 1990s really is, given the persistent federal budget deficit.   Also, while trends in environmental concerns would seem to be a plus for gas demand relative to other fuels, a growing environmentally based emphasis on conservation would tend to reduce demand. Again, no one seems quite sure where these forces will balance out.

Operational Risk.   Pipelines also face increased operational com- plexity.   Imbalances are of particular concern now that so many independent entities must be coordinated.   Pipelines are a *de facto* supplier or demander of last resort when imbalances arise, or at least are subject to complaints from all concerned despite their literal inability to control the situation.   Perhaps at some point a technological 'fix' for this problem will be available, but that solution may require a noticeable investment by pipelines.

In short, based on recent history and the current uncertainties, pipeline executives with whom we have spoken see the 1990s as a time of high risk for their industry.   It is now time to exam- ine the contrary view.

ARGUMENTS FOR LOW RISK

Conversations with those familiar with the industry, including past regulators, and our own experience suggest that three interrelated lines of reasoning underlie the perception that pipeline risk has fallen in recent years:[4]

1.   The industry's high risks are primarily due to take-or-pay disputes, which now lie mostly in the past, not the present.

2.   The industry has been compensated for whatever risks it has borne and is now bearing by an allowed rate of return equal to the cost of capital, and, moreover, the cost of capital seems to be down from its level in the mid-1980s when risks were very high.

3.   Regardless of the level of risk in the past, a combination of the shift from sales to carriage and new FERC rules either shift risk to producers or customers or make customers pay for the risks pipelines will bear in the future.

The first line of reasoning is straightforward: just as it can be said that a pitcher who has served up a bases-loaded home run is no longer in trouble, it can be argued that the pipeline industry's long nightmare is now over. The industry's massive write-

---

4   In this chapter we only describe the arguments, while Chapter 8 analyzes them. We must note that the conversations that form the basis of this section took place before comparability was a material issue.

offs are a thing of the past, according to this notion, and the high risk seen by the industry is wholly in the past.[5] The future is considerably safer.

The other two lines are addressed in detail in Chapter 3, but we next summarize them for the convenience of the reader.

Arguments that an Allowed Rate of Return Equal to the Cost of Capital Compensates for All Risks

The basic argument here is that, after all, investors are allowed a rate of return on a specified base of assets that is above the risk-free rate precisely *because* they have agreed to bear risk. They have had some losses, in fact, some very large losses, but that is just bad luck, not unfair treatment. When regulators measure the cost of capital and equate the allowed rate of return to it, investors are fully compensated for the risks they bear. In effect, this line of reasoning concludes that risks must be low because the allowed rate of return does not require a big risk premium, nor have measurements of the cost of capital in recent years shown a level of risk that is dramatically higher than that of other industries.

There are two more detailed variants of this argument and two additional pieces of evidence cited in support of it. We cover the two variants in more detail in Chapter 3, but we will summarize them here before describing the evidence cited in support.

---

[5]   See for example, Tim Kinsey (1990).   The Columbia Gas bankruptcy obviously calls this view into question.

The DCF Variant. The first variant argues that any level of future risk has already been incorporated into the cost of capital via the Discounted Cash Flow (DCF) methodology, an approach to estimation of investors' required rate of return used by many regulatory commissions.[6] The reasoning is that since the DCF method relies on the stock price, which reflects all risks, whatever risks the utility faces must be incorporated in the DCF measurement of the cost of capital.

According to this logic, regulators automatically account for regulatory risk when they equate the allowed rate of return to an estimate of the cost of capital produced by the DCF methodology. In effect, the argument is that future risks must be low because DCF estimates of the cost of capital are saying they are low.

The CAPM Variant. The other variant of the argument proceeds via exactly opposite reasoning. This version originates in the principles of modern finance. Specifically, modern finance, as embodied in the Capital Asset Pricing Model (CAPM), for example, holds that only *nondiversifiable* risks[7] affect the cost of capital. This view argues that most of the increased risks cited by the industry are fully diversifiable and therefore are irrelevant to

---

[6] See, for example, Kolbe, Read, and Hall (1984, Chapter 3) for descriptions of the DCF approach and other methods used to estimate the cost of capital.

[7] 'Nondiversifiable' risks are those that cannot be eliminated by formation of a portfolio. See, for example, Brealey and Myers (1988, Chapters 7 and 8). The topic of diversifiable versus nondiversifiable risk is discussed in Chapter 4 and Appendix A of this report.

the cost of capital.[8]   Those that do affect the cost of capital will show up when it is measured.   Since regulation should equate the allowed rate of return to the cost of capital, regulators can safely ignore any risk that does not show up when the cost of capital is measured.

In short, this line of reasoning asserts that, from the perspective of regulatory theory and practice, no adjustment need be made to the allowed rate of return for regulatory risk.[9]   Thus, the CAPM variant is quite different from the DCF variant, which says that the DCF model yields a rate of return that, because the additional risk lowers stock prices, includes an additional risk premium that is fair compensation for these risks.   Clearly, both arguments cannot be right, but as long as one or the other is right, regulators need not be too concerned: both approaches claim regulators can accept the estimated cost of capital as the fair allowed rate of return.   However, if *both* are wrong, regulators will need to reassess their current procedures.

---

[8]   Among these diversifiable risks, this logic would include the risk of regulatory disallowances of imprudently incurred costs, investments that turn out not to be 'used and useful,' the requirement to buy out take-or-pay contracts, and other past and future risks associated with uncertainty in the regulatory climate.   We continue to label these as 'regulatory risks,' without defining that term too precisely for the moment.

[9]   Representatives of the pipeline industry sometimes have a fairly straightforward retort to the CAPM variant: any method of computing the cost of capital which concludes that one can impose billions of dollars of losses on investors without affecting investors' required rate of return is seriously flawed.   If the CAPM, or any other method for computing the cost of capital, leads to different conclusions, the solution is to revise the method for computing the cost of capital.

Supporting Evidence. In addition to these two variants on the basic argument that allowed rates of return compensate pipeline investors for the risks they bear, two pieces of supporting evidence may be cited. Those who believe that pipeline risks are low may note that a number of investment projects in the natural gas industry are being funded.[10] If investors really feared excessive risks in the future, would they not pressure management to stop putting new capital into the business?[11] Moreover, those who believe pipeline risks are low might note that gas pipeline stocks in the last two or three years have been priced in excess of book value.[12] A market-to-book ratio of one is a traditional test of whether regulated investors expect to earn the cost of capital (Kolbe, Read, and Hall 1984, Chapter 2).

---

[10] It is not difficult to find, for example, various bullish Wall Street and trade press reports citing good news for gas industry stocks from time to time. See, for example, Caleb Soloman and James Tanner (1989, C12) and Caleb Soloman and Dianna Solis (1989, 1). Articles of this sort may be cited as evidence of low risk in the gas industry.

[11] Investors in regulated companies have pressured managers to retard investment in the past. An overt case is Public Service Company of Indiana whose shareholders voted overwhelmingly to "minimize future capital investment for the purpose of constructing new generating plants," until the treatment of investment in new plants improves in Indiana, because "the investments of the company shareholder should not be unreasonably put at risk through large capital programs to meet [future] demand" (*Electric Utility Week* 1985, 3).

[12] Given regulators' willingness to permit market-to-book ratios to remain below one for substantial periods in the past, those who feel pipeline risk is low today might discount the importance of market-to-book ratios, whether above or below one.

Arguments that Recent Changes Eliminate or Compensate for Pipeline Risks

The above arguments are supplemented by a third line, which appeals to recent regulatory developments.   Briefly, the argument is as follows.   Efforts by the FERC to permit some recovery of take-or-pay liabilities, changes in rate structures to permit a higher proportion of cost recovery through fixed demand charges, and adoption of gas inventory charges directly address the problems pipelines have faced.[13]   Moreover, the shift from merchant service, in which pipelines must contract to buy gas, to transportation service, in which pipelines merely ship gas for others, means pipelines are less exposed to the kinds of risks that cost them so dearly in the 1980s.   These changes result in a low level of pipeline risk, in this view.

This line does not dispute that an *integrated* gas company (i.e., one involved in every part of the industry from the wellhead to the burnertip) is indeed highly risky because natural gas has become just another commodity, the price of which responds to the fluctuations of world energy markets.   However, it does dispute that this development necessarily means higher risks for the *pipeline* segment of the industry.

The reason given is that pipelines historically shifted the risk for price and quantity changes in gas supply forward to customers via tariffs and service agreements which embodied minimum bills (obligations to pay for minimum quantities of pipeline gas) and take-or-pay contracts with gas suppliers.   While all agree that this hedge failed to insulate pipeline investors from risk, it now is

---

[13]   See Appendix B for discussion of these policies.

said to be either no longer necessary or replaced by other hedges.

For example, the switch from tariff sales to primarily transportation service is said to have fundamentally lowered the risks of the pipeline industry. According to this logic, the pipeline's merchant function (i.e., its acting as an agent to acquire gas on behalf of end users) entails substantial market risk. If most users have switched to transportation services for much of the year and perform the merchant function themselves, however, this logic concludes that there is less risk in the pipeline segment of industry than there was historically.

Even with regard to the pipelines' remaining tariff sales, pipelines now have the option of charging a Gas Inventory Charge (GIC) to recover the risks entailed in acquiring gas on customers' behalf (the gas merchant function). This new cost recovery mechanism protects pipelines from the future risks of another take-or-pay fiasco, according to this view, or at least compensates them explicitly for the risk.

A variant of this argument concerns the reliance of pipelines on a two-part rate design. As was shown in Figure 6-1, demand charges are independent of throughput and give the customer certain rights to service during peak periods. Commodity charges are paid by customers only when they take the gas. To make gas more competitive with other fuels, changes in gas pipeline rate design have shifted cost recovery to what is said to be a less risky reliance upon recovery of costs through fixed charges rather than through rate components that are sensitive to pipeline throughput. Some cite this trend as shifting risks of throughput variability from pipelines to their customers, thereby offsetting to some degree the loss of the minimum bill mechanism.

Summary

Thus the perceptions of low or reduced risk today arise from at least three perspectives.   The first states that pipeline risks are lower today than they were in the past because the take-or-pay nightmare is now over.   The second line of reasoning concludes that pipelines are compensated for the risks they face because their allowed rates of return equal their estimated costs of capital. Moreover, pipeline risks are low because that conclusion follows from the theory and empirical results of studies of the cost of capital.   Market-to-book ratios above one and continuing new investment by the industry may be cited in support of this logic. The third line of reasoning concludes that regardless of whether risk was greater or less in the past, changes in the industry and new ratemaking mechanisms either eliminate the risks by making gas producers or customers bear them, or explicitly compensate pipelines for bearing those risks.

THE RISK ANALYSIS TO COME

The problems the industry has experienced and the current dispute over the level of risk the industry faces thus suggest three major questions regarding the industry's future:

1.   Are such changes behind the pipeline indus-
     try so that from now on one can say that
     the risk exposure of the pipeline industry is
     (a) much lower than it was in the 1980s
     or (b) low on some absolute scale or (c)
     low relative to other industries?   Or do
     pipelines remain exposed to substantial risks?

2.   Are such unforeseen events and changes in the fundamental structure of the pipeline industry and its regulation reflected in the models used to estimate risk and the allowed rate of return?

3.   Does regulation now protect investors in the pipeline industry from such events by shifting risk to gas customers or suppliers, or if not, by providing compensation for bearing this risk?   Can regulation do this in the new competitive environment?

To the extent possible, we provide answers to these questions in Chapter 8.

Chapter 8
Risk Analysis for Natural Gas Pipelines

A high degree of risk is inherent in the natural gas industry's operating with a netback pricing system driven by
competition in end-user markets where there is substantial gas-on-gas and interfuel competition and where
consumption is affected by weather and business cycles.

*   *   *

The variability in prices and volumes in the end-user
sector and increasing competition for industrial fuel
markets have increased risks in the retail gas market.
[footnote omitted]   Regardless of regulatory authorization granted by PUCs to LDCs to quote flexible prices
at the burner-tip, reasonable rates of return on capital
will not be achieved by LDCs and pipelines if end-user
prices are out of line with final demand realities. Therefore, rates of return allowed by regulators for pipelines
and LDCs will have to reflect their *risk exposure, which
is greater than it has been in the past* [emphasis added]
(Hall 1987, 252, 259).

INTRODUCTION

This chapter attempts to pull the various threads from the previous chapters into a cohesive pattern. After this introduction, we consider in
several stages the various logical arguments for and against the view that
pipeline risk remains high. In the process, we present additional quanti-
tative evidence on the risk of the industry, from stock market data. At
the end, we present our views on the level of pipeline risk. This sets the
stage for the discussion in Chapter 9 on what we think are the crucial
issues that the pipeline industry and its regulators face in the 1990s.
First, however, we review the risk debate.

Traditionally, gas pipelines, as other utilities, have been treated as having
overall low risk, arising from natural monopoly and stable demand.
Regulation generally was designed to pass the remaining risk forward to
customers, who were presumed to benefit from lower required rates of
return as a result.

The 1970s and, especially, the 1980s brought the realization that gas pipelines as well as other regulated utilities could indeed be high risk industries. In fact, risk in some of these industries became so high that the traditional allocation of risk to final customers proved politically unacceptable.[1] In response to these and other forces,[2] regulatory institutions changed to shift risk away from customers back upstream to distributors, pipelines and gas producers.[3]

Despite the history of the last decade or two, however, there now is debate over the level of risk the pipeline industry faces. Chapter 7 described this debate. Briefly, the position that pipeline risk is high pointed to examples of three types of risk: regulatory, business, and financial. The position that pipeline risk is low made three arguments: the worst is in the past, pipeline investors are and have been compensated for bearing risks, and new regulatory practices and an evolving industry structure have achieved (or soon will achieve) a sustainable system in which pipelines can survive and perhaps even prosper.

This chapter analyzes these arguments in detail. The arguments are difficult to arrange in a logical order, however, because there is some intrinsic give and take among them. That is, if we could perhaps whimsically characterize the essential nature of the debate as it has come to seem to us, it is as follows:[4]

---

[1] For a review in the early 1980s, see American Gas Association (1983).

[2] In the case of pipelines, for example, producers as well as end users wanted open access transportation.

[3] As Kalt (1987, 90-91) puts it:

> As wellhead deregulation proceeds, excess demands are cleared increasingly by price adjustments, and supply and demand risks are being distributed away from end users and back upstream.

[4] For simplicity, we characterize the parties in this imaginary conversation as 'pipelines' and 'regulators,' but neither is strictly accurate. First, we use this conversation to organize the subsequent discussion, and many of the comments on both sides of that discussion are our own. Second, while regulators routinely review the logic of arguments that regulated companies put forth, and while former regulators are the source of some of our understanding of these views, we do not wish to

1.   *Pipelines*:   "It's obvious that pipelines are very risky. Look at all that has happened and all the uncertainties that remain."

      Regulators: "No, it's not obvious at all. Look at all the corrective changes that have happened already or soon will. Good times are here or just around the corner. Besides, you're paid to bear risks."

2.   *Pipelines*:   "Those responses don't stand up. No one can be sure the final regulatory system will let us survive in a world where the Commission is actively promoting competition, too, or what the economic future holds. The new regulatory mechanisms in place to date are incomplete and far from proven, still more major changes are proposed, and it is likely to take a long time for final mechanisms to be determined. Given the risk, we're not sure the return is compensatory."

      Regulators: "Well, if things don't work exactly right the first time, we'll fix them, at least to the extent you can convince us they really need to be fixed. As for the

---

imply that all regulators are skeptical of the view that pipelines are risky, nor that regulators see themselves as pipeline critics rather than as the neutral arbiters of the various views, from pipelines and others, brought before them.

return, when you stop making
new investments, then we'll
start to worry."

Our discussion essentially follows this hypothetical conversation. First
we review the factors that lead pipelines to conclude that their high risk
is obvious. Then we examine the responses to those arguments, testing
them as we do to see if they are truly responsive. The final comment
made by our 'regulators,' however, is in part nonresponsive and in part
more fundamental than any of the others. The first sentence is nonresponsive, since one of the major risks perceived by the pipeline industry is uncertainty over what the rules will actually be.

The second sentence, however, seems to us to be truly fundamental: if
pipelines are still making new investments, why does the Federal Energy
Regulatory Commission (FERC) need to worry about whether risk compensation is adequate -- it obviously is, else pipelines would not sink new
capital in the business. We use this comment in Chapter 9 as a springboard to raise what we feel are issues that are far more important than
the precise level of risk the pipeline industry faces. We find that the
issue of whether risk compensation is adequate is not obvious at all.

"IT'S OBVIOUS THAT PIPELINES ARE VERY RISKY"

The history and changing nature of the pipeline industry are discussed in
Chapter 6 and Appendix B. From the business person's view of risk, the
gas pipeline industry has certainly been associated in recent years with
extreme risk. The 1980s produced dramatic changes in natural gas markets and regulatory institutions for transmission of natural gas via interstate pipelines. Wellhead gas prices were partially, then fully, deregulated, declining gas demand and collapsing energy prices created a massive
'take-or-pay' liability for natural gas pipelines, and the FERC undertook
steps that effectively opened access to the interstate system for competing
suppliers of natural gas. The result is a growing switch from the traditional pipeline role of merchant (a delivered price combining both gas
supply and transmission services) to or toward pure transportation of
third-party gas. The result also has been poor earnings and weak financial performance generally, as documented at the start of Chapter 6.

Future Risks Identified by Pipelines

Asked to itemize future risks, as opposed to past ones, the following seemed to be considered most important by at least some firms. (We add the risk category into which they fall in parentheses.)

1. Sensitivity to world energy markets (business);

2. Volatility of gas supply and resolution of contract disputes (business/regulatory);

3. Environmental cleanup costs (business/regulatory);[5]

4. Ability to meet peak demand requirements in a world where customers can 'cherry pick' among service alternatives (business/regulatory);

5. The lack of historical experience in providing a mix of transportation and tariff gas services (i.e., risks of cost recovery) (business/regulatory);

6. The threat of deregulation, or a return to more comprehensive regulation, in the event of a new energy crisis (regulatory);

7. Conservation response or a carbon tax in response to global warming (business/regulatory);

8. Bypass of pipelines by LDCs and new entrants, and the need for discounting (business);

9. Loss of system control (operational);

---

[5]   For example, Panhandle Eastern's estimate of potential PCB liabilities is reported by *The Wall Street Journal* (1991a) to have increased by $330 million since its 1989 acquisition of Texas Eastern Corporation.

10. The requirement to serve without an adequate cost recovery mechanism for being the 'supplier of last resort' (i.e., the inability to abandon service, particularly merchant service) (regulatory);

11. Conflicts between federal and state regulation that would prohibit cost recovery (regulatory);

12. Proposed legislation to increase competition for pipelines' *transportation* function (business/regulatory);

13. Additional competition for pipelines' *merchant* function via the mega-NOPR (business/regulatory); and

14. Gas quality control risks (operational).

Regulatory/Business Risks

Generically, managers tend to characterize business/regulatory risks as emerging from government actions (e.g., environmental cleanup costs) and the interaction of regulatory and competitive constraints. Examples are given above and in Chapter 7.

Discounting is a good example of the kinds of concerns pipeline managers express. Current FERC policy allows pipelines considerable latitude in setting transportation rates. The result is widespread discounting, as shown in Chapter 6. The dollar value of the revenue dilution is considerable. Pipelines who are exposed to discounting of rates fear that a 'just and reasonable' rate (i.e., one that permits them to earn their cost of capital) will form a ceiling for rates, which are in turn forced downward by competition.

Initial policy positions by the FERC recognized that making pipelines fully responsible for the revenue dilution from discounting would be counterproductive. Sometime afterward, however, the FERC proposed ratemaking methodologies that would in effect make pipelines 'eat' the revenues lost from discounting, by assuming that all transportation occurred at the maximum rate. It would be difficult to imagine a system more guaranteed to impose asymmetric losses on pipelines (INGAA 1988).

The FERC (1989) appeared to have recognized this point in its rate design policy statement in May 1989.[6] However, even after that statement was issued, the policy seemed to be one of dealing with the issue on a 'case-by-case' approach. Pipelines complained that as late as early 1991, FERC staff input to actual FERC decisions reflected the old approach, which guarantees a shortfall from the allowed rate of return if all other cost-of-service parameters are estimated precisely. Very recent evidence suggests this may finally have changed, but the existence of the problem for any period highlights just how risky and uncertain the regulatory rules can be.

More generally, discounting is a response to competition. Competition brings its own set of challenges and risks to regulated companies. This is a topic we discuss in more detail at the end of this chapter.

Another example, not raised previously, is that pipelines also express concerns about lawsuits based on the antitrust laws. They fear customers may demand access to pipeline transportation services when pipeline gas acquisition costs exceed competitive levels (via the 'essential facility' doctrine) and to tariff gas at rolled-in prices when the opposite is true (via the 'price squeeze' doctrine).

'Comparability' is the fastest growing concern. The FERC's mega-NOPR actively addresses whether rule changes are needed to put the pipeline's merchant function on a comparable basis to that of other gas sellers, by having pipelines 'unbundle' the various components of their merchant function (e.g., storage) and through other changes. Uncertainty over the ultimate rules is an obvious source of risk. Moreover, under the mega-NOPR pipelines are apparently to share in the transition costs of such a change, akin to the 'equitable sharing' of take-or-pay contract losses.

---

[6]   See Appendix A.

We understand that a number of potential transition costs exist. For example, many pipelines have finished restructuring their take-or-pay contracts, but those contracts still may prove off-target if comparability comes, requiring further restructuring. Also, relatively new, high-(regulatory)-cost gathering line investments might be stranded if competition due to comparability prevents rolled-in pricing of gathering line costs. In view of recent experience, pipelines are understandably concerned at such prospects.

More generally, comparability is only one example of the overall uncertainty about the ultimate regulatory system and about whether it will truly be compatible with the growing level of competition. The entire regulatory process has become more complex in more recent periods. For example, the list of *appearances* by attorneys for interested parties in the AGD-I decision[7] before the D.C. Court of Appeals ran to 11 pages! Everything seems to take forever, with more people disputing more issues. As Figure 6-9 showed, one result has been a notable increase in regulatory lag. Pipelines simply are uncertain about whether costs will be recovered in such an environment.

Operational Risks

Increasing operational risks are inherent in the growing gas transportation function. These are difficult to quantify in terms of the traditional measures of risk, but they have the potential to affect pipelines' 'bottom line.' These risks arise from the fact that the operational and accounting complexity of a system involving thousands of individual transactions is much greater than one where the pipeline management is the central clearinghouse and decision maker acting as the gas merchant on behalf of all buyers and sellers. The growth in the number of transactions and the attendant increase in imbalances was depicted in Chapter 6.

From an operational point of view, the change experienced by pipelines is akin to the difference between a railroad, where a central dispatch coordinates movements and prevents conflicts, and a highway, where every auto and truck owner simply acts in his or her own interest. As

---

[7]   See Appendix A.

any driver stuck in traffic knows, the result is not necessarily the most efficient use of the system.

From a financial and accounting point of view, the question of gas imbalances may be even more frightening for pipeline managers. Imagine a bank in which large numbers of people have the right to withdraw large quantities of money without the bank discovering either the amounts or the precise people who withdrew the cash until long afterwards. Imagine that customers do so with the promise that other persons, unknown to the pipelines, will deposit equal amounts on the customers' behalf, and that these offsets are done through hundreds of bilateral agreements which are unknown to the bank. Imagine also that individual participants routinely fail to keep their side of these unknown agreements and that pipelines have no way of monitoring the performance of their counterparts in the separate transaction on a timely basis. Imagine finally that there is no commonly accepted method of forcing all these checkbooks to balance or to deter 'gaming the system.' This is the essence of the transportation imbalances problem on gas pipelines.

When the pipeline was the primary provider of gas merchant services, it coordinated the inputs and outflows to achieve the 'gas balance,' i.e., the need to coordinate overall supply and demand. The balancing was at an aggregate level, with no need to match deliveries to certain customers with receipts of gas from specific suppliers. As a transporter, those supply and demand decisions are highly disaggregated as individual buyers acquire gas under their own contracts. The number of transactions has gone up enormously and gas balancing is now transaction-specific instead of aggregate.

At least two major problems may arise from these disaggregated decisions. First, buyers and sellers wish to have flexibility in their receipt and delivery points.[8] But freely making these decisions on a disaggregate basis may ignore pipeline capacity surpluses and bottlenecks in the operating system. Far-flung gas systems, in particular, may have geo-

---

[8]   This is one element of the emerging debate on 'comparability,' discussed in more detail below. In that debate, shippers want to be sure that pipelines do not use the imbalance problem as an excuse *unnecessarily* to restrict comparable access to the system.

graphic imbalances and bottlenecks that impair aggregate balancing. Pipeline gas movements must now be undertaken so as not to discriminate against specific disaggregated transactions. Unless individual participants get the correct price signals, an impossibility without real-time monitoring and pricing, the results will underperform relative to centralized dispatch.

Secondly, the pipeline becomes the residual buyer or seller in case of the default on the part of individual buyers and sellers. Indeed such 'imbalances' are virtually inevitable due to the inability of the pipeline to detect an imbalance soon enough to correct it.

These disaggregate imbalances have an as yet undetermined effect on pipeline risks. One issue is the effect of such numerous bilateral arrangements on pipeline throughput and efficient routing of gas movements. Clearly the administrative costs of accounting for gas flows (whose gas flowed on a particular day?), financial obligations (who owes whom?), and system management have increased. In this complex system, pipelines fear that widespread litigation may ultimately prove necessary to resolve transportation contract disputes.

Pipelines also express concern that such imbalances can be used as a substitute for firm sales service during peak periods, an incentive that is encouraged by low financial penalties for imbalances and long make-up periods to correct imbalances. This incentive can become a serious problem if the entire system exhibits an aggregate imbalance during a peak period, which is highly likely. In effect, there is a 'run on the bank,' where all accounts are motivated to exercise overdraft privileges.[9] Under the pipeline's sales service obligation, the pipeline is bound to have the inventory of gas to honor such aggregate imbalances up to its contract demand level for firm sales.

The present system also can create incentives for financially opportunistic behavior. Shippers may be motivated to take high-cost gas from the pipeline during peak periods and replace it later during the off-peaks with

---

[9]   Recall the experience during the December 1989 cold spell, cited in Chapter 6, in which pipelines were owed more than twice the amount of gas they owed others.

cheaper gas, for example. Or financially troubled shippers may be tempted to overrun their accounts in hopes of staving off bankruptcy.[10]

Also, all gas is not perfectly fungible. Where certain end users require gas free of certain impurities, the pipeline must insure that the average quality of gas in the pipeline meets that standard. A producer that injects low-quality gas may not be offset by a countervailing imbalance in another account of higher quality gas. Pipelines are at risk under their tariffs for gas quality.

At a minimum, the current regulatory uncertainty surrounding all of these imbalance issues is unsettling. The system for accounting for imbalances could produce a continuing financial dilution for pipelines. Presumably this could be handled with a cost-of-service item akin to an allowance for bad debts, which would provide explicit payments for imbalance losses in the tariff. However, no such mechanism is in place at present.

More troublesome would be the emergence of acute imbalances arising from broader economic forces, such as a run on the pipeline's gas supply during an energy crisis via imbalances, or conversely, forcing the pipeline to absorb surplus gas during plant shutdowns, for example. These could make the pipeline industry subject to more nondiversifiable risk.

Conclusion

In sum, there seem to be plenty of things for pipelines to worry about. Given these problems, it is not surprising that their initial view is something like "It's obvious that we're very risky." Now it is time to consider the second and third parts of our imaginary conversation, the first response by our 'regulators,' and whether counterarguments to that response invalidate it.

---

[10] R. Skip Horvath (1989a, 21-23) notes that system operation and accounting procedures are unable to respond in a timely way, which remains true today in almost all cases. Imbalances are detected two to three months after they occur, and participants dispute whether the reconciliation should be in kind or in cash and who is responsible for correcting the imbalance.

"NO, IT'S NOT OBVIOUS AT ALL," AND
"THOSE RESPONSES DON'T STAND UP"

We sketched the responses to the argument that pipeline risk is obviously high above and discussed them in Chapter 7. Now it is time to analyze the details of those arguments using the economic tool kit provided in Chapter 4. Do these responses imply that pipelines' concern about their current risk is unfounded?

Is the Nightmare of the 1980s Now Over?

The first response to the view that pipelines are obviously risky is that pipeline risk is in the past. The take-or-pay problem undoubtedly caused major losses, and may yet cause some pipelines more losses, but, it is said, nothing of that magnitude is now on the horizon for the industry generally. The past is unfortunate, but regulators only need worry about compensation for future risks, not past risks.

Choice of a Baseline. Implicit in the above argument is that the 1980s are the baseline against which risk is to be measured. However, the 1980s are potentially a misleading baseline for this purpose. Nobody denies that a pipeline that acquires no gas is less risky in some risk dimensions than one operating under the 1980s rules for recovery of the cost of gas supply. The latter can, at best, recover without profit its expenditures for gas supply, but can suffer a massive loss if such supplies are acquired at above market prices under take-or-pay contracts. Its gas purchases are a clear case of 'Heads I break even, tails I lose.'

However, a reduction in one major risk does not logically imply that the remaining risks are 'low.' In any discussion of risk, a major practical issue is the baseline against which current risks in the gas pipeline industry are measured. Figure 8-1 illustrates the point with a hypothetical graph of pipeline risk over time. Suppose that overall risk of the pipeline industry can be measured accurately by some single index (the particular index need not be identified[11]). The period of the 1960s and early 1970s is drawn as an essentially stable period where the market for natural gas (when available at regulated prices) was assured. The chief risk was inadequate supply (the other side of the coin of the assured demand). The figure shows risks rising during the 1970s as natural gas shortages and curtailments reduced throughput and exposed pipelines to curtailment liabilities.

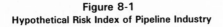

Figure 8-1
Hypothetical Risk Index of Pipeline Industry

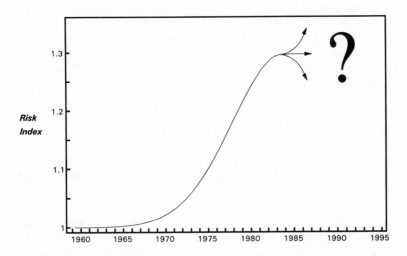

With the passage of the Natural Gas Policy Act in 1978, the industry entered a new era in which new gas supply could command market prices. Pipelines contracted for extensive supplies of gas at what turned out to be very high prices. The risk to pipelines that gas would be priced too high seemed to be hedged, however, because minimum bill provisions required gas distribution companies to pay pipelines for gas whether it

---

[11]   The last part of Chapter 4 would suggest that the fair allowed rate of return in an 'everything turns out right' outcome might be a candidate for the index, since this figure depends directly on the probabilities and magnitudes of downside disasters.

could be sold or not. As long as the minimum bill hedge held, pipelines were shielded from any downside risk of take-or-pay obligations if the gas were nonmarketable. But hidden in the new environment was a substantial increase in risks for pipeline investors. The latent risks inherent in such a regulatory scheme became manifest when the FERC approved Order 380 (on recovery of minimum bills). Subsequent key FERC decisions, such as Orders 436 and 500 (encouraging open access transportation), are at least in part responses to the risk shift created by Order 380.

It is unclear how overall risk has evolved since the mid-1980s, which is one reason this study was commissioned. Those who believe pipeline risks are now modest focus on the reduced level of take-or-pay risk as compared with the recent past. However, the lower levels of risk relative to the mid- to late-1980s may still mean risk is quite high compared to pre-NGPA levels, or even compared to U.S. industry generally. And even though take-or-pay risk has decreased, logically it might be true that other risks have increased by a more than offsetting amount.

This report addresses candidates for such increased risks below. For example, 'comparability' issues (especially transition costs) and the increasing emphasis on competition generally, combined with a tight regulatory constraint, can leave pipelines very vulnerable. Thus, the fact that take-or-pay risk is down is not by itself dispositive of the level of pipeline risk today.

In practice, the issue of the appropriate baseline becomes significant when the immediate past becomes a baseline for measurement of both risk and the necessary rate of return. Risk and allowed rate of return are presumed to be in equilibrium during this period; the argument for a reduction in the allowed rate of return is then presumed to follow from the conclusion of reduced risk of a specific variety. This conclusion obviously does not follow if risks were not taken into account adequately in the rate of return during the relevant base period, or if offsetting increases in risk arising from other factors are not properly accounted for. These issues are discussed explicitly below.

First, however, we examine the available empirical evidence on the level of pipeline risk over time.

Market Risk Evidence. Chapter 4 stressed the basic importance of non-diversifiable risk to the cost of capital. Appendix A notes that the best single measure of nondiversifiable risk, and therefore the best single measure of the risk of a stock, is known as 'beta.' Beta measures the extent to which the stock responds to fluctuations in the market. Stocks with average risk respond on average about as the market does, rising or falling by 10 percent when the market rises or falls by 10 percent. Such stocks have a beta of 1.0 by definition. Stocks with betas above one exaggerate the swings in the market, while those with betas below one understate the swings in the market.

Figure 8-2 plots the betas for a sample of publicly traded gas pipeline common stocks.[12] Actually, there are two lines on the figure: one for the full sample, and one for the four pipelines identified in Chapter 6 as the nearest thing to 'pure plays' in the gas pipeline business in 1990.[13] The 'pure plays' display noticeably higher risk during the 1980s. These betas are quite high, indicating substantial risk, but the pattern may appear somewhat surprising. The betas peak in the early 1980s, when the risk that led to the losses of the mid- to late-1980s was still potential rather than realized. Indeed, the betas fall as the bad news becomes manifest. Moreover, the equity betas at the end of the period for the larger sample are lower than those in 1977, although the betas for the near-'pure plays' are higher.[14] It is also worth noting that the climb of

---

[12] The companies are Arkla, Coastal, Columbia, CNG, KN Energy, Panhandle, Sonat, Texas Eastern, Transco, and Williams. Data were not available for all companies in all years, but all available data were averaged for each year. The betas are from Merrill Lynch's *Security Risk Evaluation* service as of December of each year.

[13] Those four are Panhandle, Sonat, Transco, and Williams. Again, data for some years for some of these pipelines are missing. Also, we did not attempt to determine whether these four were the closest to pure plays over the whole sample period.

[14] Actually, care is required in reading these numbers. They are the standard beta calculation, which uses five years of monthly data. Thus, the '1977' numbers are for January 1973 through December 1977, and the '1990' numbers are for January 1986 through December 1990. Thus, the earliest data points include the period of the first oil price shock and the resulting recession, while the next points (for 1979) include the second oil price shock. Data for 1978, which provides the cleanest

equity betas in the early 1980s is sharper than the climb in allowed rates of return on equity depicted in Figure 6-1, reproduced here for convenience as Figure 8-3.[15]

Figure 8-2
Equity Betas for Gas Pipeline Companies

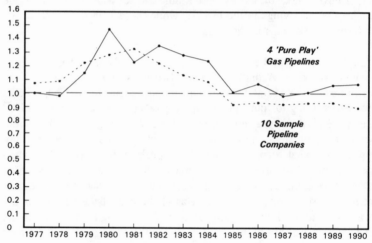

Source: Merrill Lynch Security Risk Evaluation.

In actuality, it is not surprising that the betas are not high during the period of take-or-pay losses. There is little reason to believe the take-or-pay losses are highly correlated with macroeconomic cycles, so there is little reason to expect betas would be heavily affected.

It may be surprising, however, that betas are not higher at the end of the period than at the start. Attempting to explain just why betas move is

'pre-NGPA' comparison, indicate the lowest beta for the four pure plays over the entire period.

15   Recall that the figure compares the rates of return on equity approved by the Commission to the actual rates of return the companies earned. The implication here is that with climbing betas and greater risk, the allowed rates of return should have climbed, too, all else equal.

Figure 8-3
Allowed and Actual Rates of Return on Equity

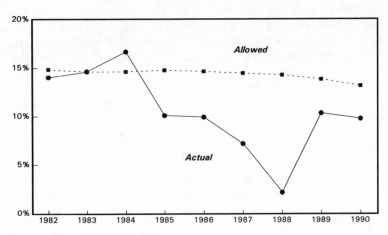

Sources: Allowed ROE: INGAA Risk Survey.
           Actual ROE: Net Utility Operating Income Survey.

intrinsically speculative. Still, there is one obvious reason for the relative difference between standing and ending values in Figure 8-2. Realized bad news in the form of Orders 380 and 436 and the take-or-pay
losses that ultimately resulted hit the gas pipeline industry just as the
stock market was taking off for one of its best periods ever. To the
extent that the bad news for pipelines was due to good news for the
economy (i.e., low energy prices), one might argue that the drop in betas
truly reflects the pipelines' nondiversifiable risk. However, we think it
more likely that a substantial part of the bad news for the pipeline industry was a diversifiable (albeit asymmetric) risk, in which case the fall in
betas is due to pipelines' having an abnormal proportion of 'bad outcomes' at a time the market was rising. In that case, part of the fall is
due to a data problem and 'true' pipeline betas may have been above the
measured betas.[16]

---

[16]   Other pipeline industry observers suggest that while the rise in betas is due to
       recognition of the potential for trouble as high-priced pipeline gas faced falling spot
       market prices, the fall in betas reflects the resolution of that uncertainty, albeit at

It is worth pausing at this stage to examine what the betas we do have imply about pipeline financial performance. In Chapter 6, we raised the possibility that the below-average performance of the pipeline samples was because they were stocks of below-average risk. Figure 8-2 shows the reverse is true for the pure plays. Thus, if the pure play pipeline sample had performed as predicted by the CAPM, they would have done *better* than the market in the 1980s. The predicted results for the two samples are depicted in Figures 8-4 and 8-5. The two figures reproduce from Figure 6-3 the stock market and gas pipeline total stock returns for the ten-company and four-company samples, respectively. Each figure adds a third line, the predicted gas pipeline sample stock return based on the betas shown in Figure 8-2. The pure play sample of gas pipeline companies underperformed even more when their relative risk is taken into account.[17]

Figure 8-5, the purest look at the pipeline business itself, is particularly striking. From 1980 to 1990, someone who invested only in one-month Treasury bills would have seen $1.00 grow to $2.27.[18] Pipeline investors ended up with only $1.51.

To return to the analysis of the betas themselves, a third possible reason for their surprising pattern is also a more general concern: the betas in Figure 8-2 are potentially misleading indicators of relative overall risk over time for the pipeline business, because they do not control for differences in capital structure. As discussed in Appendix A, the addition of debt adds financial risk also, which increases beta. If the average capital structure of the industry has changed over time, that change could be responsible for some of the pattern we see in Figure 8-2.

---

a large cost to the pipeline industry, and the view that the Commission now had matters under control. There is no way to know for sure whether this view or the view in the text better explains the fall.

[17]   And if the true pipeline betas are higher than the estimated ones, as just hypothesized, pipeline underperformance is greater still.

[18]   See Ibbotson Associates (1991). This calculation, like Figures 8-4 and 8-5, ignores taxes and transaction costs.

Figure 8-4
Predicted vs. Actual Returns for 10 Sample Pipeline Companies

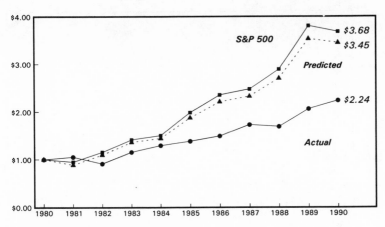

Source: Value Line, Merrill Lynch, Ibbotson Associates, Inc.

To explore this possibility, Figure 8-6 plots the weighted-average debt and equity betas for the two groups shown in Figure 8-2.[19] The peak remains in the early 1980s, as before, but the picture is somewhat more consistent with Figure 8-1 in that recent overall betas remain above those of the pre-NGPA period. Thus, part of the apparent decline in betas at the end of the period is indeed explained by capital structure changes. The total nondiversifiable business risk of the pipeline industry is higher

---

19   The capital structure data are from the *Value Line Investment Survey*, various dates, and debt betas were calculated from the corporate debt series in the Ibbotson Associates' Stocks, Bonds, Bills and Inflation database, 1990 edition. The debt betas used were 0.25 for 1977-1980 and 0.35 for 1981 and beyond. Capital structure was calculated using the year-end book value of debt and the year-end market value of equity. This may tend to overstate the market-value proportion of debt somewhat, particularly in the early years of the sample when embedded interest rates tended to be lower than market rates, which would tend to slightly understate the weighted-average betas in the early years.

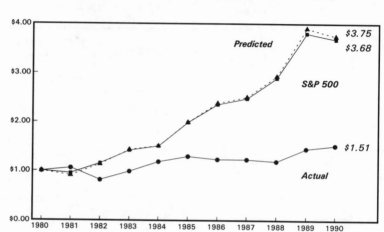

Figure 8-5
Predicted vs. Actual Returns for 4 'Pure Play' Gas Pipelines

Source: Value Line, Merrill Lynch, Ibbotson Associates, Inc.

than in pre-NGPA days and trended generally upward in the second half of the 1980s, after rising and falling sharply in the early 1980s.[20]

Another theme of Chapter 4 is that some asymmetric risks that do not affect beta still affect the rate of return regulated companies require. Such risks, *if realized*, should affect the total volatility of the stock. Figure 8-7 plots the basic measure of total stock volatility, the 'standard deviation' of the rate of return on the stock, for the two sample groups in Figure 8-2.[21] Note that the standard deviations in Figure 8-7 are of the order of 10 percent *per month*, plus or minus. Because there is approximately a two-thirds chance that a randomly selected point from

---

[20]   Again, if the true betas are underestimated in recent years, the upward trend in the late 1980s would be more pronounced.

[21]   The standard deviations of the stocks' rates of return are also known as 'sigmas.' These data are derived from the Merrill Lynch regressions that yield the betas reported in Figure 8-1. Thus, they too are for the 60 months ending with December of the year shown.

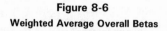

Figure 8-6

Weighted Average Overall Betas

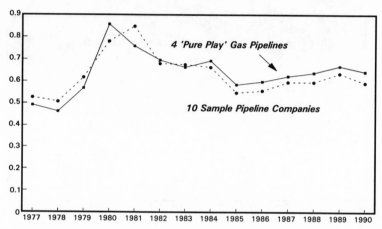

Source: Value Line, Ibbotson Associates, Inc.

a distribution will fall within one standard deviation of the mean, investors in gas pipeline stocks had nearly one chance in three of either gaining or losing more than 10 percent of the value of their investment in any given month. We lack data on the average standard deviation of common stocks, but the stock market as a whole (which is not strictly comparable because it reflects maximum risk reduction through diversification) has a standard deviation of a bit under 6 percent per month.

Note also that the standard deviations stay roughly level for both the 'pure plays' and the broader sample of companies with more revenues from other lines of business. There is no sign of a dramatic fall in total risk in recent years.[22]

---

[22]   The fact that standard deviation is roughly level while overall betas in the late 1970s are lower suggests more of the pipelines' risk was diversifiable in the late 1970s.

Figure 8-7

Equity Monthly Standard Deviation

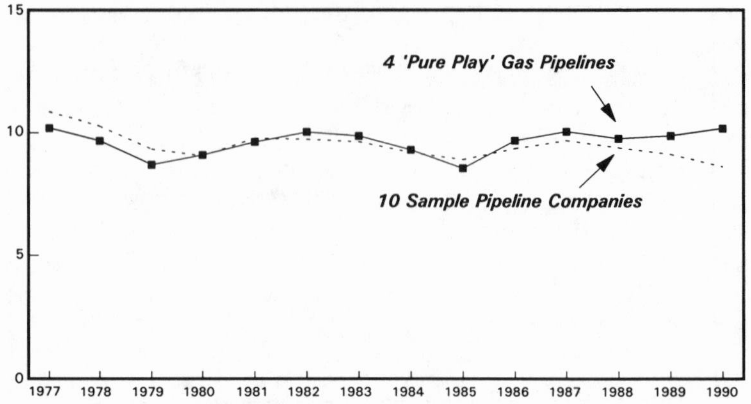

4 'Pure Play' Gas Pipelines

10 Sample Pipeline Companies

Source: Merrill Lynch Security Risk Evaluation.

In sum, the statistical risk evidence thus tends to confirm the general intuition about the pattern of pipeline risk over time, but more is needed here than reference to statistical measures of risk. Much of the pipeline risk debate is conducted on logical grounds, and statistical evidence that conflicted with what someone believed true on logical grounds might well be disregarded. It is time to move on to the logical arguments.

Have Pipelines Been Compensated for the Risks They Bear?

The second response to the argument that pipeline risks are obviously high is that pipelines have been compensated for bearing those risks through the allowed rate of return on their investment. Moreover, the cost of capital seems to be down from its high levels in the 1980s, when take-or-pay exposure was at its maximum. If the cost of capital measurement techniques give estimates in the mid-teens as one regulated company is absorbing a large take-or-pay loss, or as another regulated company is writing off half its book value for a canceled nuclear plant, regulators may be excused for wondering just how risky these companies can be once such extraordinary losses are absorbed.[23]

In addition to the basic argument that an allowed rate of return equal to the cost of capital provides adequate compensation, supporting evidence in the form of market-to-book ratios and continuing pipeline investment may be cited. We defer discussion of the last of these points to the end of this chapter, but we cover the others now.

An Allowed Rate of Return Equal to the Cost of Capital. Much of this response and the counterarguments to it have been covered in previous chapters. The basic point that an allowed rate of return equal to the cost of capital provides adequate compensation has become automatic, in fact, too much so.[24] When it is challenged, 'proofs' of its validity come forth. Thus, Chapter 3 discusses the 'DCF' and 'CAPM' defenses of the practice. Briefly, the DCF variant holds that since the stock price reflects all risks, and since the DCF method of estimating the cost of capital relies on the stock price, the resulting cost of capital estimate will compensate for all risks. The CAPM variant holds that since only non-diversifiable risk affects the cost of capital, regulators may safely ignore most or all regulatory risks such as take-or-pay losses.

Chapter 4, however, shows that both variants are false. Just as pharmaceutical companies require high rates of return on successful drugs to pay for the failures, and oil companies require high rates of return on successful wells to pay for dry holes, and computer software companies require high rates of return on successful packages to pay for the flops,

---

[23] Kinsey (1990) found that a "downturn in pipeline financial performance [occurred] in the latter part of the decade. . . . In every performance category, with the exception of total debt as a percent of total capital, the industry experienced a decline in credit quality in the 1985-89 period."

[24] For example, Kolbe, Read and Hall (1984) has a section in Chapter 6 entitled, "Why the Allowed Rate of Return Should Equal the Cost of Capital." The notion of asymmetric risk and its consequences is never mentioned. Instead, the book simply assumes that other cost-of-service elements in the company's revenue requirement are set so they are expected to cover all other costs. When this condition fails, however, so does the prescription to equate the allowed rate of return to the cost of capital.

so too do regulated companies require high rates of return on investments that become used and useful to pay for the large losses that asymmetric regulation permits.

The cost of capital is a breakeven rate of return, and the average of breaking even and incurring a loss is a loss. The goal that truly permits companies to break even is an *overall* rate of return, on successful and unsuccessful investments *combined*, that investors expect will equal the cost of capital. That goal cannot possibly be achieved by a policy that restricts the return on successful investments to the cost of capital, but permits returns well below that level in other circumstances.

Thus one simply cannot proceed from the result of a relatively low cost of capital computed via the standard methodologies to the conclusion either that pipeline risks are low or that pipeline investors have been compensated for the risks they bear. Even if done without error, the standard cost of capital estimation methodologies by their very nature exclude considerations of the types of asymmetric risks that can cause the largest losses.

For example, take-or-pay risks are definitely asymmetric, because pipelines earn nothing on the gas if it can be sold, since gas costs are passed through to pipeline customers without any markup, but bear large losses if it cannot. The take-or-pay asymmetry means allowed rates of return have been far too low to provide fair compensation during the recent era of take-or-pay losses. (Recall Figure 8-3, which shows no sign of an increase in allowed rates of return during the period of heavy take-or-pay losses.)

In short, regulatory perceptions based on cost of capital estimates can vastly understate the actual level of risk regulated companies face. Indeed, this argument for low risk can become circular. Biased estimates of the necessary allowed rate of return can become a rationale for ignoring the problem, rather than for questioning the process itself. Thus, we must proceed to an issue by issue analysis of the pipeline industry which specifically addresses the presence or absence of regulatory or other asymmetric risk. First, however, we examine evidence that might in principle be cited for or against the view that pipeline risks are low: the companies' market-to-book ratios.

Supporting Evidence: Market-to-Book Ratios. A standard theorem of rate regulation is that if a company consists entirely of assets subject to rate regulation, and if investors expect to earn their cost of capital on the rate base value of those assets, the market-to-rate-base ratio will be 1.0.[25] In original cost rate base jurisdictions, this translates approximately into a market-to-book ratio of 1.0. This theorem needs to be applied with caution, and, as discussed below, is clearly not as robust as once thought. Still, the evidence should be examined.

Figure 8-8 plots the market-to-book ratios of the two groups of pipelines in the earlier figures, over the same period. Figure 8-9 plots the same information for the logarithm of the market-to-book ratio, which provides a more accurate visual comparison.[26] The ratios of the 'pure plays' lie below the larger sample's ratios for most of the period.

The data suggest companies that were in trouble in the early 1980s, just as the risk measures of the earlier figures peaked, but which now appear headed for better times. The apparent rebound in the mid-1980s, however, may be due in part to conservative provisions for take-or-pay liabilities, which would have lowered pipeline book values excessively.

Several cautions are needed about the interpretation of these data, some specific to gas pipelines and some more general. One specific concern is the pipelines' declining rate bases. (Recall Figure 6-5.) The market-to-book ratio is infinite if the book value is zero, so a shrinking rate base will tend to lead to high market-to-book ratios. Also, if a pipeline has had a large take-or-pay loss on its books but its other operations are unaffected, valuable assets that led to a modest premium over book value before the loss could show up as a large premium after the loss has

---

[25]   See, for example, Kolbe and Read and Hall (1984, Chapter 6).

[26]   The market-to-book ratio can fall no further than zero but can rise indefinitely, which gives an exaggerated visual importance to the values above one in Figure 8-8. For example, if you lose 50 percent of your investment, you need to earn 100 percent on the next investment to make up the difference. Since minus 50 percent and plus 100 percent, in either order, leave you where you started, they are of equal importance monetarily. The logarithmic scale of Figure 8-9 provides visual symmetry to shortages and overages. (Note that the logarithm of one is zero, so values below one in Figure 8-8 are negative in Figure 8-9, and conversely.)

Figure 8-8
Market-to-Book Ratios for Natural Gas Pipelines

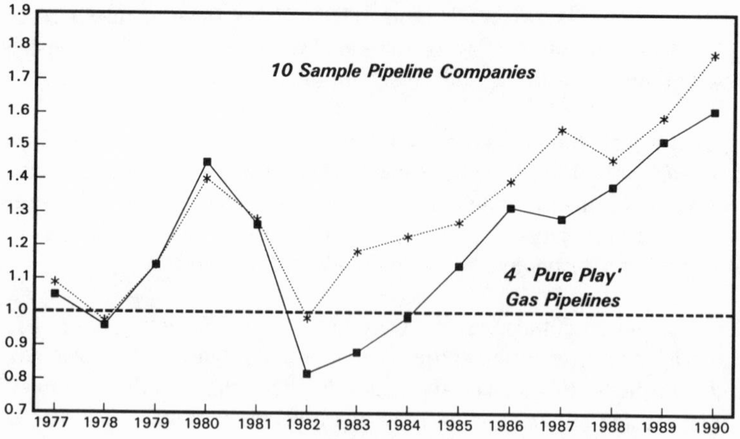

Source: Value Line, January 4, 1991.

reduced book value.[27] Finally, some pipelines' parents also own gas reserves, which may be booked at values far below current market value.

However, other, more general concerns about interpretation of market-to-book ratios exist, too. Quite apart from the numerical results themselves, market-to-book ratio evidence needs to be treated with caution, for several reasons.

The first is that regulation in practice is 'sticky.' Rate cases only occur from time to time, which implies expected rates of return may exceed or fall short of the cost of capital between hearings even if the expected rate of return were exactly equal to the cost of capital when the rates were

---

[27]   The booking of take-or-pay losses in general may lead to biases in either direction. Some companies may book the undiscounted expected loss, which is an overestimate of the market value of the loss. This will lead to artificially high market-to-book ratios. Other companies may book nothing, on the grounds that the amount cannot be reliably quantified, which could lead to artificially low market-to-book ratios.

Figure 8-9
Natural Log of Market-to-Book Ratios

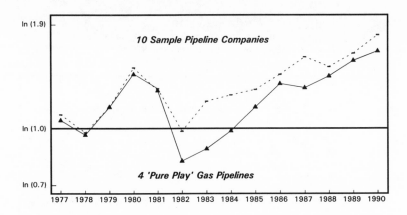

Source: Value Line, January 4, 1991.

set. Moreover, regulators appear to adjust the allowed rate of return for changes in the cost of capital with a lag. Ordinarily there is a wide range of cost of capital estimates in a hearing, and regulators therefore have a record that gives them wide latitude. They appear to avoid rapid changes from the rate of return in the last hearing for a particular type of company.

This problem is illustrated in Figure 8-10. The figure reproduces the allowed rate of return from Figure 6-1, adds two interest rate series, annual U.S. Treasury short-term bill and long-term bond yields, and adds a capital asset pricing model estimate of the cost of capital.[28] The allowed returns clearly do *not* track changes in interest rates closely, nor

---

[28]   See Chapter 3 for discussion of the CAPM. The CAPM estimate in Figure 8-10 equals the short-term Treasury rate in the figure plus the 'pure play' beta from Figure 8-2 times the usual estimate of the market risk premium, 8.5 percent. This is a standard application of the most widely used method of estimating the cost of capital (except in rate regulation, which still uses the less robust DCF approach more often than the CAPM).

do they typically equal the cost of capital. Allowed returns in rate regulation are simply much more stable than capital markets are. Moreover, with the exception of 1987, the allowed rates of return lie below the cost of capital of the 'pure play' sample.

Figure 8-10

Pipeline Allowed Returns and `Pure Play' Cost of Capital vs. T-Bill and
Long-Term Government Bond Yields

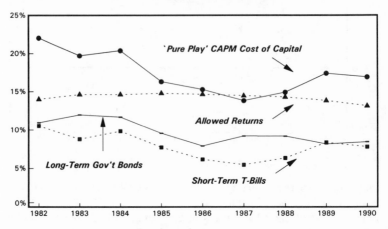

Sources: INGAA Risk Survey, Merrill Lynch, Ibbotson Associates, Inc.

The 'stickiness' of regulation and the behavior of regulators mean that allowed rates of return will lag behind the cost of capital when it is rising or falling. The market-to-book ratio can fluctuate around one in a long cycle in such circumstances, equaling one only on average, even if there are no other concerns about the market-to-book ratio at all. A market-to-book ratio above one during a part of the cycle thus may still mean a regulated company's expected rate of return falls short of the cost of capital on average.

The second basic reason that market-to-book ratios need to be treated with caution is that most regulated companies are not 'pure plays' in the regulated line of business. In particular, market-to-book ratios in unregulated businesses may be far above one even when the company in question just expects to earn its cost of capital.[29] The reason is that book values do not generally correspond to market values. Inflation and the difference between book and economic depreciation are responsible. It is only rate regulation, which bases earnings on the net book value of regulated assets, that gives market-to-book ratios for regulated companies any meaning. Since even the 'pure play' sample in Figures 8-6 and 8-7 is not really a sample of literal pure plays, this problem may affect both samples.

Third, and perhaps most fundamental, financial economics is no longer as sure of the economic meaning of market values as it once was. The October 1987 stock market crash, for example, is hard to square with the notion that stock market value always measures fundamental economic value. It is likely that *relative* market prices are right, but we can no longer be confident that *absolute* market prices are right, except on average over some period of time.[30] However, the market-to-book ratio test for regulated companies is a test of *absolute* value. If regulated companies' absolute economic value were measured precisely by their stock prices and if relative values among all stocks were also right, then the rest of the market's absolute economic values would be measured precisely, too, contrary to the evidence of the October crash. Thus, the crash leads us to question the merits of the market-to-book ratio test, also.

In short, market-to-book ratios may be indicative of expected rates of return above or below the cost of capital over a long period of time, at least for 'pure play' companies regulated on a net book value rate base, but the evidence is not very precise even for these companies. As a practical matter, market-to-book ratios for regulated companies are a good deal less useful a test of investor expectations than was once thought.

---

[29] This may be one reason the larger sample shows higher market-to-book ratios in Figures 8-8 and 8-9.

[30] There have been many articles on the October crash, many of which explore it in great detail. The comments made here originate in Myers (1988, 8-9).

Will Changes in the Pipeline Business and the Regulatory Rules that Govern It Make Pipelines Low-Risk Companies?

Thus, the broad-brush responses (i.e., that pipelines' nightmare is over and that regulation automatically compensates pipelines adequately by equating the allowed rate of return to the cost of capital) do not by themselves dispose of the risk debate. We are left with the hard task of evaluating the changes one by one, to see to what extent they make pipelines into low-risk companies.

A recent decision by the FERC gives tangible expression to the view that changes in the industry reduce pipeline risk.[31] First the Commission compared risks today with those of 1985, during the midst of the take-or-pay crisis for the industry:

> We believe that the risks faced by Northern Border in 1990 are comparable to the risks faced by Northern Border in 1985 when the Commission accepted the settlement return on equity of 14.5 percent. Therefore, a 3.7 point risk premium above the current yield for 30-year Treasury bonds would result in a return on equity of 12.0 percent, rather than 13 percent.[32]

Next the Commission considered the fact that reliance upon cost recovery through demand charges affected pipeline risk:

> In examining the risk factors to which Northern Border is exposed, we must first give consideration to Northern Border's cost-of-service tariff. Because all costs are recovered through demand charges, cost recovery is

---

[31]   The pipeline in question, Northern Border, is atypical, since it is a transportation-only, cost-of-service regulated investment. But we lack published decisions with clear discussions of the issues in other cases (since most cases settle). Moreover, while the circumstances of this pipeline differ from most in the industry, the *themes* seem common to the FERC's view of the entire industry.

[32]   Federal Energy Regulatory Commission, *Northern Border Pipeline Company*, Docket Nos. RP89-33-000, CP78-124-014, Order Approving Contested Settlement with Modification and Certificate Amendment, Issued July 30, 1990.

assured, regardless of the volumes transported. The tariff also includes provisions to 'true up' costs and revenues in the event they do not match. These are unusual tariff provisions which few pipelines have. Most pipelines recover only a portion of their costs in a demand charge and have no true-up provisions. Thus, Northern Border faces considerably less risk than the ordinary pipeline.[33]

And finally, the Commission relied upon the fact that the pipeline at issue does not perform a merchant function:

Also we cannot ignore the fact that Northern Border is a transportation-only pipeline. Thus, unlike pipelines which perform a merchant function, Northern Border has no risks whatsoever associated with the sales or marketing of its gas, and as a result it has not had to face the possibility of take-or-pay liability.[34]

The FERC's decision confirms that Northern Border tends to be more completely affected by the changing nature of the pipeline business than other pipelines. Still, the decision process was not unique. It proceeded by defining the mid-1980s as a baseline and argued for a reduced rate of return based on a lower cost of capital now as compared with then. It cited shifts to cost recovery through demand charges and provision of transportation rather than merchant services as additional factors arguing for a low return. The only argument not raised was that Gas Inventory Charges (GICs) would compensate for the risk of the merchant function, because Northern Border has no merchant function.

There are thus three questions to address in this section:

1.   Does the change in rate design, to shift cost recovery to demand charges, make pipelines low-risk companies?

---

[33]   *Ibid.*

[34]   *Ibid.*

2.   Does the shift from the merchant function (sales)
        to the transportation function (carriage) make
        pipelines low-risk companies?

3.   Do GICs or a market-based pricing mechanism
        compensate for the risks of the merchant func-
        tion?

The Effect of Changes in Pipeline Rate Design on Pipeline Risk. The
risk a pipeline faces is in part a function of rate design. Where service
is 'firm,' rates are generally composed of a demand charge and a commodity charge (see Figures 6-1 and 6-2). As discussed in Chapter 6, a
demand charge is basically a capacity charge that reserves the right to
service during peak periods and is paid regardless of the level of service
actually rendered. The commodity charge is based on the quantity of gas
actually provided to the customer. As noted in Chapter 6 (recall Figure
6-15), pipeline rate designs have changed over the years, allocating first
more, then less fixed-cost recovery to the demand charge.

Actual throughput will generally vary from planned throughput. The
extent of these variances probably will depend partly on nondiversifiable
factors (e.g., general business conditions, which affect demand for gas
by end users) and partly on diversifiable factors (e.g., the number of
really cold days in December). Either way, all else equal, the pipeline
is less exposed to risk of throughput variances with a higher proportion
of fixed costs in the demand charge (at least to the extent that costs are
in fact recovered in the demand charge). The shift of costs to the demand charge as rate design moved from the 'United' to the 'Modified
Fixed/Variable' formulas (see Chapter 6) thus is sometimes said to have
reduced pipeline risk substantially.

Of course, no one argues that the MFV formula eliminates risk entirely.
An important goal in the design of the MFV methodology was to load
enough fixed cost recovery (including all return on equity and related
costs) onto the commodity charge in order to put risk back onto the
pipeline. The finding was that otherwise pipelines would have little

incentive to maximize throughput (and thus economic efficiency).[35] The converse of strong motivation to succeed is strong medicine if one fails.

Still, it is logical that if a lower proportion of fixed cost recovery depends on volume, pipeline risk is lower, all else equal. When examined closely, however, the argument for low risk in the gas pipeline industry due to changes in rate design founders on the key assumption, 'all else equal.' Consider the following:

1. The shift in rate design was a market necessity compelled by the fact that prior regulatory policies had made much gas unsellable; the marketing environment under today's rate structures is much tougher than when the Seaboard and United methodologies were employed.

2. Much of pipelines' FERC-approved volume, and even more of their actual volume, is shifting to interruptible transportation, which is subject to heavy and growing discounting and which has only one-part volumetric rates (i.e., no fixed demand charges).

3. Pipelines no longer have minimum bills for shifting risk to customers.

4. The ability to impose demand charges is not the same as a throughput commitment that binds the customer.

5. Any benefits in the shift of costs are often more than offset by adjustments to other ratemaking parameters.

---

[35] The FERC in its *Policy Statement Providing Guidance with Respect to the Designing of Rates* noted that pipelines must not be exposed to the risks of rate variances through discounting if they are to be motivated to use this as a tool to maximize throughput. See Appendix A.

We consider these one by one.

Changed Market Conditions. As discussed in Chapter 6, the shift in rate structures itself was a response to changed market conditions which also meant an increase in total risk of the integrated pipeline system from wellhead to burnertip. For a pipeline with adequate and secure reserves purchased at below-market prices, the problem historically was to ration excess demand to the available supply. The switch in rate structures came in large degree because the gas commodity became overpriced under the old rate structure. A pipeline attempting to sell or transport gas facing heavy competition at the end use, even under a rate structure where a lower portion of fixed costs is recovered in the commodity charge, is not necessarily less risky than one able to move all the cheap gas it can get under a rate structure where the commodity charge recovers a higher portion of the fixed costs. Indeed, the normal presumption would be the reverse.

Shifts in the Composition of Demand toward Interruptible Service. The argument for lower risk addresses the allocation of the portion of the costs assigned to the firm service (between demand and commodity),[36] but does not address the increasing share of the costs being allocated to interruptible transportation service. The share of fixed costs that must be recovered from the interruptible service has increased from 3 percent to 28 percent more recently.[37]

This shift in responsibility for cost recovery is a direct consequence of the shift in volumes from sales to, primarily, *interruptible* transportation. Indeed, 64 percent of the transported volumes have been interruptible in recent periods (INGAA 1991a). Interruptible transportation rates are one-part volumetric rates which are intended to recover variable costs and some fixed costs, but which do not include payment of a fixed de-

---

[36]   Recall Figure 6-19.

[37]   Recall Figure 6-20.

mand charge. Moreover, they are increasingly subject to discounting.[38] Thus, shifts in the composition of demand tend to offset shifts in the rate structure.

Loss of Minimum Bill Mechanism. Another way in which 'all else' is *not* equal is with regard to contracts with pipeline customers. Pipelines in recent periods do not have the hedge of minimum bills available to protect their sales in end-use markets. Nor can pipelines easily relieve themselves of the obligation to supply gas when regulated firm sales contracts expire. Simply put, the MFV methodology has been implemented during a period where total system risk is up and where the previous set of regulatory and contractual rules to shift those risks forward to customers no longer is in place.[39]

Demand Charges Are Not Throughput Agreements. The ability to impose demand charges is not the same thing as a throughput agreement that contractually binds the customer to pay the fixed cost over the life of the facility.[40] First, some pipelines recently have been forced to discount demand charges because of competition. Moreover, regulators may yet change the rules to facilitate efforts by customers to avoid commitments to pay for sunk capacity.[41]

---

[38] Recall Figure 6-21.

[39] Gas Inventory Charges are intended to deal with some of the risks that arise from the end of minimum bills. We consider GICs below.

[40] For example, owners of baseball teams should not assume that they have eliminated risk by giving fans the option to buy season tickets, even if that guarantees a certain proportion of seats will be paid for that season. The owner has to worry about keeping ticket sales high season after season, and fans may decide to get their entertainment elsewhere if the local baseball team fades.

[41] Note that the *FERC Policy Statement Providing Guidance with Respect to the Designing of Rates* affirmed the observation that if price incentives were designed to ration peak capacity more efficiently among customers, then those customers must be free to make capacity adjustments. But if this is the case, pipelines cannot permanently shift risk forward to customers via demand charges. The 'mega-NOPR' provides a forum where this risk may yet be realized.

Offsetting Adjustments to Other Ratemaking Parameters. Pipelines also complain that all else is not equal with regard to other ratemaking parameters. For example, regulators may establish the throughput 'bogey' for ratemaking at higher levels under the MFV methodology than previously. Since the pipeline will earn its cost of capital only if the projected throughput is reached (again all else being equal), any reduction in risk from changes in rate design may be more than fully offset by raising the throughput hurdle. Note that the risk of throughput variances is decidedly asymmetric (i.e., throughput cannot exceed 100 percent on a sustained basis, and downtime from scheduled and unscheduled maintenance guarantees it will be less on average).

Straight Fixed-Variable Rate Design Under Comparability. The mega-NOPR proposes to shift from MFV to a straight fixed-variable rate design. By itself, this again would tend to reduce pipeline risks, all else equal. It need hardly be added that all else is decidedly not equal under the proposal in the mega-NOPR. Moreover, the risk of the portion of fixed costs recovered from interruptible rates would be unaffected.[42]

Conclusion. Thus, 'all else' is decidedly not equal in rate design. The shift to more fixed cost in the demand charge has been accompanied by, and indeed is in large part a response to, other changes that increase pipeline risk markedly. The conclusion that rate design changes make pipeline risk low does not hold up.

The Effect of the Shift from Sales to Transportation on Pipeline Risk. When we speak to industry observers and former regulators, we sometimes get the impression that they believe pure gas transportation service is very low risk. The notion that pipeline gas sales are more risky than gas transportation has at least two bases. One is that since open access

---

[42]   Also, a reduction in risk has value to pipelines only if it does not lead to a corresponding reduction in expected rates of return. In fact, FERC Staff have proposed cuts in the allowed rates of return on equity if a straight fixed-variable rate design is adapted (Knight 1991).

has been implemented, there simply are far more competitors for sales than for transportation. The second basis is that transportation is believed not to have the downside potential that take-or-pay clauses gave sales.

The greater number of competitors for sales is indisputable. But looking at the history of the industry, this reflects more of an increase in the risk of sales than a decrease in the risk of transportation. The same parties suggest that unlike its policy in the past, the FERC now is disinclined to let the certification process protect existing pipelines from new competitors or new competitors from their own misjudgments of demand.[43] If so, the risk of *both* sales and transportation is up, the former just more than the latter.

As far as take-or-pay losses go, we believe there can be no disagreement that a pipeline cost recovery mechanism that sets rates for tariff gas (i.e., pipeline-owned gas) equal to its cost, yet leaves pipelines exposed to substantial downside exposure for unmarketable gas purchased under take-or-pay contracts, entails substantial risks. Thus it seems logical that anything that reduces pipeline exposure to such risks must reduce pipeline risk overall. Presumably, if customers rather than pipelines are performing the gas merchant function, pipeline risks of cost under-recovery are reduced.[44] The view that the shift to transportation makes pipelines low-risk companies may seem the logical next step.[45]

---

[43] For example, in *Transwestern Pipeline Company,* Docket No. CP90-2294 (issued January 17, 1991), the Commission approved construction under Section 7(c) without a showing of firm contracts and supporting market data for the entire capacity of the pipeline. However, it did put the pipeline at risk for cost recovery to the extent that all of the new capacity was not subscribed under firm contracts at the time of its initial rate filing. See, also, Arkla Energy Resources, Docket No. CP89-2174 Order Issuing Certificate (issued Jan. 17, 1991), pp. 17-18; *ANR Pipeline Company,* Docket Nos. CP89-637, *et al.,* Preliminary Determination and Order Issuing Certificates and Terminating Dockets (issued Jan. 17, 1991), pp. 34-35.

[44] As noted above, however, comparability could lead to a new round of losses during a transition.

[45] Recall the earlier quotation from the Northern Border decision.

Stated this way, and given the discussion to this point in the chapter, it is immediately clear that the final step goes too far. First, it is important to know just what is meant by 'the shift to transportation.' Second, a reduction from the level of risk associated with resolution of take-or-pay problems does not imply that the remaining risks are 'low.'[46] Finally, there may be other factors that tend to offset the reduction in exposure to take-or-pay risk.

The Switch to Transportation as Misnomer. The 'switch to transportation' is not uniform, which has important consequences for pipeline risk exposure. Figures 6-10 to 6-13 showed that pipelines have indeed largely become transportation rather than merchant carriers. During 1990, transportation averaged 79 percent of total throughout for end users, local distribution companies, and marketers. If carriage on behalf of other pipelines is included, transportation represented 82 percent of total throughput in 1990, up from 3 percent in 1982 (INGAA 1991a, Tables A-1, A-2).

To understand these data, it is first important to recognize that there have in fact been fewer *conversions* from firm sales service to firm transportation service than from sales to transportation.[47] Based on three years' worth of data, 22 percent of total maximum daily quantity has been converted relative to the 1988 base year. The main definitional problem for the view that the 'switch to transportation' reduces pipeline risk, however, is that the switch from sales to transportation is much greater on a volumetric basis than on a capacity basis. Many customers are relying on sales service during peak days and switching to transportation in the off-peak. (Recall Figures 6-14 to 6-16.) Furthermore, use of

---

[46] Nor does a reduction in risk from the era of take-or-pay problems imply that pipeline allowed rates of return can be reduced, unless those rates of return were compensatory at the time. (And as we discussed above, those rates of return were *not* compensatory for take-or-pay risks.)

[47] To the extent that these conversions are a permanent abandonment of firm sales service, the pipelines will ultimately be able to reduce their supply commitments accordingly. However, as discussed in Appendix A, pipelines have a good deal less flexibility to abandon service than their customers have to reduce their obligations to pipelines.

pipeline sales service has become more spiked even within the peak season. Figure 6-17 showed that the peak day represents an increasing magnitude relative to the average demand during the peak season. Increasingly, pipeline sales capacity is being relied upon for the brief surge of the winter season.

For these portions of pipeline capacity, the obligation to serve has not gone away. Rather, demand has become more spiked in the peak winter period as the pipeline cannot sell its gas in the off-peak. In these circumstances, the shift to transportation combined with the obligation to serve leaves pipelines with a difficult gas purchasing problem.

Moreover, the fact that this switch is largely seasonal suggests that it may well not be driven by economic efficiency considerations. Instead, it seems likely that the pipelines have not been able to structure their contracts or tariffs to make commodity prices for sales match spot market levels. Part of the problem may lie in Order 451, which gives producers the initiative in restructuring gas contracts. Obviously, a producer with more high-cost contracts (whose price would probably come down in a renegotiation) than low-cost contracts (whose price would probably rise) will simply not open negotiation. Thus pipeline gas costs on old contracts will tend to rise over time, making off-peak sales harder. At the same time, the requirement to charge the weighted-average cost of gas on-peak implies a pipeline's peak sales price will tend to be below spot prices on peak.

Possible solutions to this problem, in principle, include pricing correctly the peak period sales function via seasonal rates and correctly implemented Gas Inventory Charges. Whatever the most appropriate solution, the fact that customers find pipeline sales preferable during the peak suggests that pipelines may not have been properly compensated for providing peak period service.[48]

---

[48] This 'cherry-picking' of peak-period service is particularly troubling to pipelines because they have a continuing obligation to the sales service even beyond the contract period, as discussed in Appendix A. Of course, this problem would continue even if the seasonal distortions ended.

Comparability might eventually reduce this risk, to the extent pipelines can abandon their obligation to serve and sell gas at a profit adequate to compensate for long-term contracting risks. The problem, of course, is in the transition. Pipelines *now* stand ready to serve their entire firm sales commitments. Competition for those sales may trigger another round of losses from contract renegotiations.

Moreover, pipelines have *not* been compensated for the risk of such losses. First, absent GICs, pipelines sell gas at cost, with no profit at all. Second, GICs are not uniformly in place. Third, even GICs in place have not been operating anywhere nearly long enough to provide an adequate 'insurance premium' for the potential losses under comparability.[49] Finally, the bulk of the GICs now in place can hardly have been designed with comparability transition risks in mind, given how rapidly the issue has moved into active consideration.

Thus, the further 'switch to transportation' associated with comparability clearly increases pipeline risks, at least in some important dimensions.

'All Other Things' Are Not Equal. The shift to transportation from tariff sales raises much the same set of issues as the effect of rate restructuring on pipeline risk. A pipeline supplying transportation services only is not necessarily less risky than one supplying tariff sales only, with its risk position hedged via sales of underpriced gas under minimum bills. Moreover, to the extent that the FERC is in fact less inclined than formerly to let the certification process protect pipelines from competition, or is more inclined to set higher throughput requirements, transportation risk is up on an absolute scale as well.[50] Total risk is up for the transportation-only pipeline, and it is clearly no longer as squarely on the shoulders of the customer.

---

[49]   INGAA (1991b, 7) estimates a range from $1.4 billion to $7.7 billion in supply-related comparability costs. That range might be narrower now that the details of the mega-NOPR are available, but the potential for very large losses is clear.

[50]   For example, the FERC may increase the projected throughput on a pipeline to a level that the pipeline maintains is an economic and engineering impossibility.

Other Offsetting Factors. The switch to transportation, even properly understood, brings new risks as well. The operational problems associated with coordination of many contracts to which the pipeline is not a party, discussed above, are one example. Moreover, even a pure transportation pipeline is not necessarily low risk. Oil pipelines have been common carriers, owning none of the goods they ship, for decades, but oil pipelines are an extremely risky business.

Figures 8-11, 8-12, and 8-13 compare the volatility of oil and gas pipeline rates of return over two periods, 1977-1984 and 1977-1988. (Figure 8-13 is only for the gas pipeline subsidiaries of the parent companies, and so provides a 'pure play' to 'pure play' comparison with oil pipelines.) Two facts stand out.

Figure 8-11
Accounting Standard Deviation of Returns on Assets

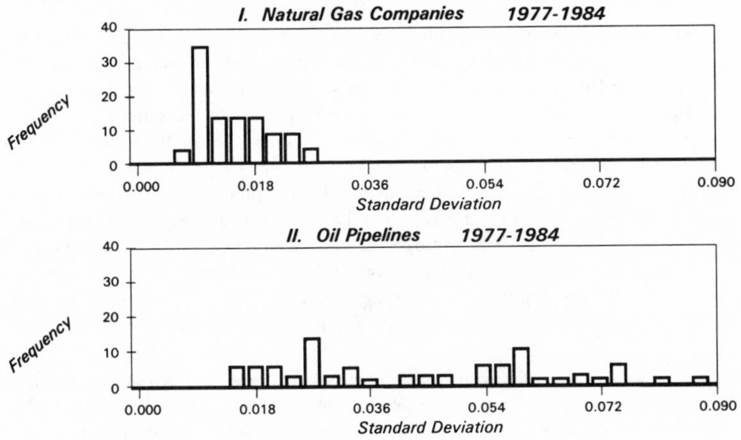

I. Natural Gas Companies   *1977-1984*

II. Oil Pipelines   *1977-1984*

* 6 oil pipeline companies have sigmas greater than 0.099

First, oil pipelines display a vastly greater range of earnings volatility than gas pipelines historically, even though oil pipelines are transportation-only investments. Second, the changes from Figure 8-11 to Figures 8-12 and 8-13 clearly imply that the volatility of gas pipeline earnings increased in the 1985-1988 period. We suspect take-or-pay problems are the chief source of this increase, but we cannot be sure that some of it is not also due to increased risks for both sales and transportation. Based on what we know of both industries, we believe oil pipelines to be riskier than gas pipelines as they are regulated at present, but oil pipelines prove beyond a doubt that pure transportation service by pipeline is not intrinsically low risk.[51] In fact, the reverse is true: the intrinsic risk of the pipeline transportation business is substantial.[52]

In sum, as the certification process for gas pipelines eases, and as end-use competition continues to grow, even transportation-only pipelines can face substantial risk.

---

[51] The chief source of oil pipeline risk is competition for a service requiring large sunk investments up front. No Commission-approved certificate of public convenience and necessity is required to build an oil pipeline, a protection that gas pipelines may effectively have lost in recent years. At the same time, there are probably more sources of competition for oil pipelines since the material they ship is easier to handle. For example, trucks, ships, and barges provide some competition for oil pipelines. Also, someone may build a refinery at the far end of a crude oil pipeline or next to water access and put a petroleum products pipeline out of business. There have been large oil pipeline projects built that sat empty for several early years or failed entirely. Whether this level of competition could come to the gas industry if it were regulated as loosely as oil pipelines historically were is not clear, but something like it is not out of the question.

[52] The basic source of risk for pipeline transportation is the need to sink substantial costs in advance to build the pipeline. Oil pipelines' parent companies typically hedge oil pipeline earnings for some of these risks with 'throughput and deficiency agreements,' which assure the pipeline subsidiary will at least be able to service its debt. Since new oil pipelines were typically financed with 90 percent or more debt before 1985, these guarantees could prove substantial.

Demand charges serve the same function for gas pipelines. As noted above, however, demand charges are not guaranteed for the life of the pipeline, as discounting of demand charges demonstrates. The level of and rationale for demand charges has changed several times over the years, in response to other changes in the industry's circumstances. Thus, demand charges by themselves cannot remove the intrinsic risk that large sunk costs create.

Figure 8-12
Accounting Standard Deviation of Returns on Assets

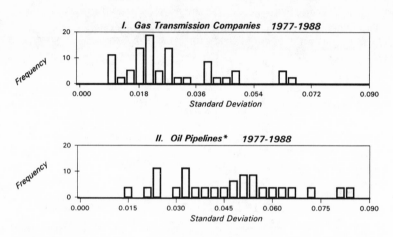

Sample from Compustat   * 6 oil pipeline companies have standard deviations >0.099

Gas Inventory Charges and Risk Allocation[53]

When pipelines (or anyone else) sign long-term contracts to purchase gas, they bear risk. Pipelines pass gas costs through to customers without markup, so they receive no compensation for that risk. Historically, this was reasonable because minimum bills passed the risk through to the pipeline's customers. However, the forces that led to Order 380 and its offshoots eliminated the minimum bill hedge and left pipelines bearing substantial costs for which they had never been compensated.

Pipelines today still have sales obligations to some customers, which pipelines cannot unilaterally abandon. Thus, to the extent they have long-term contracts to meet those obligations, pipelines bear risk for

---

[53] We are indebted to Frank Graves and George Hall for long conversations regarding GICs, although this discussion is of course our own interpretation. This discussion was written before comparability made GICs potentially obsolete. However, GICs will be a factor at least until comparability comes, and the principles involved in the present discussion have relevance even under comparability.

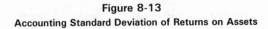

Figure 8-13
Accounting Standard Deviation of Returns on Assets

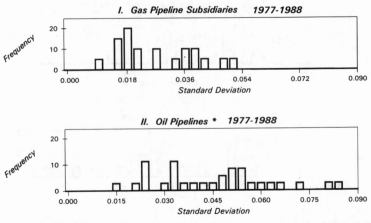

Sample from I/B/E/S.   * 6 oil pipeline companies have standard deviations greater than 0.099

which they are not compensated. The intention of GICs is to provide such compensation.

However, our conversations have left us believing that there are two views of how GICs should do so. In one view, GICs are a replacement for minimum bills, to be entered into by pipeline customers with full knowledge of the risks involved. In this case, GICs are less compensation for risk than another attempt to pass the risk through to customers. In the other view, GICs are an insurance payment pipelines should get to bear the risk that gas contracts needed to meet service obligations will prove in excess of what the market will support. These two views are loosely correlated with the difference between 'cost based' and 'market based' GICs, but we have not seen formal acknowledgment that a cost-based GIC is supposed to be the equivalent of a minimum bill.

The view of GICs as minimum bills, if implemented *and enforced*, in principle would relieve pipelines of contracting risks. In practice, it is hard to believe such a mechanism could be enforced under all conditions. First, it is economically desirable for some risk to remain with the pipelines, to encourage efficient contracting practices, and this effect would show up in prudence reviews, if nowhere else. Second, if contract and spot prices got far enough out of line again, it is likely that the same forces that led to pipelines' bearing some of the cost last time could make pipelines bear some of the cost next time.[54]

Thus, we are convinced that GICs must at least to some degree serve the insurance function. In this role, GICs in principle are a possible solution to the remaining asymmetric risk of the merchant function.[55] However, identification of the right answer in principle does not mean the answer will turn out to be right in practice. The difficulty with GICs so far is in their implementation, to which we now turn.

The FERC recognized as early as Order No. 436 that freely opening up the gas pipeline industry to transportation and third-party contracting involved a decided asymmetry. Incentives for expedient behavior would arise if customers could freely 'shop around' during periods when spot prices were below contract prices and shift back to tariff sales when the converse was true. Other customers or pipeline investors would bear take-or-pay obligations in the first case and the high incremental costs of selling additional output at rolled in prices in the second case. Many felt that there was a natural efficiency advantage in letting the pipeline perform the gas acquisition function (scale advantages in matching aggregate demand and supply and efficient use of pipeline routes and services), and this would be lost if customers were given a free option to call upon a pipeline's gas supply under contract while 'playing the market' for spot gas supplies.

---

[54]   It is not clear that pipelines have as much political clout as either end-use customers and gas producers, for example.

[55]   Recall that in Chapter 4 the two remedies for asymmetric risk were an increase in the allowed rate of return or an explicit payment to bear the risk, akin to an insurance premium. If the insurance premium were set at the right level, pipeline investors would willingly bear the asymmetric risk, just as insurance companies willingly sell insurance policies. Indeed, since the risk of gas purchase contracting has nothing to do with the size of the rate base, an explicit compensation mechanism such as a properly computed GIC that is based on the risk itself is clearly superior to an adjustment to the allowed rate of return on the rate base.

While FERC's objectives in imposing GICs thus have corresponded to the risks pipelines face, the implementation has been slow, incomplete, and filled with remaining uncertainties.[56] FERC's policy statement on gas inventory charges was not published until two years after Order No. 436 and three years after Order 380, and as of 1991, many pipelines do not have GICs in place. Those that do have GICs have provided an extremely risky service with no explicit cost-recovery mechanisms for a considerable period of time. The remainder with no GIC mechanism continue in this status.[57] Moreover, there remains considerable uncer-

---

[56]   As Richard J. Pierce, Jr. (1990, 9) put it:

> At present, most distributors can choose whether to buy gas from their traditional pipeline suppliers or from other suppliers. Even if a distributor chooses to rely entirely on other sources of supply, however, the pipeline remains legally obligated to sell the distributor the volume of gas specified in the certificate of public convenience and necessity governing the relationship between the pipeline and the distributor under the Natural Gas Act of 1938.

> This situation is not viable in the competitive wholesale gas market that FERC has created. With distributors free to buy no gas from a pipeline, the pipeline has no incentive to enter into contracts to purchase enough gas to meet the full certificated requirements of its distributors. The combination of a pipeline's freedom to buy no gas from a pipeline creates a classic free-rider problem. If allowed to persist, it will end the way all free-rider situations end -- the bus will not be there when the riders try to climb aboard. *Pipelines are not being compensated for the cost of maintaining an inventory of gas under contract* [emphasis added].

[57]   According to Arthur K Worldwide Organization/Cambridge Energy Research Associates (1991, 66-67),

> [t]he institution of GICs by pipelines has been slow in coming, in part because GIC settlements must give the customers a viable alternative to system supply -- specifically the ability to convert their supply entitlement to a "comparable" firm transportation right. There has been some progress toward instituting GICs in the last year, with several pipelines accepting FERC-approved GICs in late 1990.

> The implemented GICs as of December 31, 1990 were:

> - Columbia Gas Transmission
> - Natural Gas Pipeline

> - Texas Eastern Transmission
> - Transcontinental Gas

tainty about GICs that have been approved. GICs have been remanded by the court, proved insufficient to recover gas costs, or have had no volumes nominated.

The chief sticking point in implementing the GIC mechanism has been the imposition of an insurance concept (without a reliable claims experience) onto an industry regulated via traditional cost-of-service concepts. After all, how does one establish a 'cost-of-service' concept for earthquake insurance in say, Washington, D.C., where there have been no claims in recent history?[58]

Undoubtedly, this is why there is an emerging tendency for the GIC to differ in practice from the statements of principle. The conflict in practice arises from a discrepancy between a system ostensibly designed to compensate pipelines for bearing risk while responding to concerns that GICs could become a profit center! Of course, if the insured risk does not occur, insurance companies *should* pocket the 'profits,' as compensation for having borne the risk. Thus there are times that GICs *should* be a profit center, and regulations designed to guarantee that this never occurs will equally guarantee that GICs are *not* compensatory.[59]

- Northern Natural Gas   •   Transwestern Pipeline
- Northwest Pipeline

[58]   As originally proposed, the FERC would be required to grapple with an exercise equivalent to computing the 'just and reasonable' price for an index option for the S&P 500 in the Chicago options pits. As Pierce (1990, 10) puts it:

   A gas inventory charge is analogous to an option contract in an unregulated market. Prices of option contracts tend to be highly volatile, with large short-term price variations triggered by market participants' changing expectations with respect to future conditions in the market for the commodity subject to the option contract. The administrative rigidity of regulation renders it a poor vehicle for governing a volatile market.

[59]   If GICs were set up as full replacements for minimum bills, of course, this issue in principle would not arise. However, one GIC has apparently been in place with customers paying the GIC but not taking the gas. Unless the GIC includes the full cost to the pipeline of the gas not taken, a possibility that may be hard to square with the fact that the pipeline's customers have not taken the gas, this GIC has *not* served as a full replacement for minimum bills, and the pipeline in question *has*

Another problem is that under FERC policy, customers have the ability to change their nominations for GIC service at relatively short notice, say annually.[60] Unlike investments in pipeline capacity, however, pipelines can in principle adjust the vintage of their gas supply contracts to match the duration of their customer obligations.[61] This means, of course, that the system that could emerge might prove a far cry from one where a pipeline acts as an agent for customers in acquiring long-term gas reserves at an agreed-upon price: a pipeline that contracted for gas supplies for any period longer than its customers are committed to pay a GIC would expose its shareholders to substantial risk, if gas prices fall sharply. Absent a long-term GIC insurance mechanism, the optimal strategy would seem to be a series of short-term contracts, which pass the risk of gas price fluctuations back to customers in a different way than minimum bills did.

On the other hand, if gas prices go up sharply, pipeline customers may complain that pipelines 'imprudently' failed to sign long-term contracts. If the FERC were to respond favorably to such a claim, the result would be a new round of asymmetric risk. Thus, unless the Commission convincingly signals its acceptance of the economic consequences of the incentives its regulations create, pipeline investors will still face asymmetric risks, GICs or no.[62]

Finally, the difficulties in GIC implementation are all the more troublesome because virtually every analysis of this industry has concluded that the disasters of the last two decades have arisen from the Supreme Court's *Phillips* decision that required the FERC (then the Federal Power

---

borne contracting risk in trying to meet its service obligation. Other difficulties have arisen with other GICs in place.

[60]   The Policy Statement contemplates that customers would renominate their service requirements at regular intervals.

[61]   However, note that the pipeline must announce a firm price or pricing formula before knowing the demand for the service, thereby incurring the risk of contracting for an amount different from that actually demanded.

[62]   Of course, even if this Commission were fully willing to accept these consequences, pipelines remain at risk of future commissions having a different view.

Commission) to regulate wellhead prices (Pierce 1988, 8). Virtually every regulatory reform initiative has started with the need to conform wellhead prices to burnertip competitive realities. Yet if one looks at the original intent of the GIC policy statement, regulators were once again contemplating procedures for setting the price of the insurance premium on the gas commodity using 'cost-of-service' principles. If the commodity itself cannot be regulated sensibly using traditional cost of service principles, what reason is there to believe that an *option* to buy that commodity is suited to such regulation?

Perhaps in recognition of these problems, we understand that the FERC has recently been concentrating on 'market-based' GICs, to be approved as part of a pipeline's agreement to sales 'comparability.' This can provide little or no comfort to the industry, however, for two reasons.

First, a pipeline would have to agree to a comparability system that is likely to pose additional risks, and perhaps substantial additional risks,[63] before it can be compensated for the gas purchasing risks it *already* bears. Second, even a free market in GICs is likely to take time, and perhaps a long time, to equilibrate at the right price. Insurance policies are hard to price in the absence of sound actuarial data, and such data on the future gas-sales risks facing pipelines are likely to be highly speculative. Finally, GICs themselves may be rendered obsolete by comparability but without adequate compensation to pipelines for the transition costs.

In sum, FERC's stated intentions with regard to GICs are explicit efforts to correct past practices that create asymmetric risk, and so address the issues raised in Chapter 4. To the extent the reality falls short, however, pipelines will continue to bear asymmetric risks with inadequate or only implicit compensation mechanisms.

---

[63] Recall the pipeline survey estimates of $1.4 billion to $7.7 billion for implementing comparability, cited earlier. Pipelines are expected to absorb at least a portion of any transition costs caused by the mega-NOPR.

CONCLUSIONS ON THE LEVEL OF RISK

Our analysis to this point leads to findings in two areas: the risks of regulated industries generally, and the risks of the pipeline industry specifically. Although some of the former were developed in Chapter 4, they are important to our conclusions about the current risk of the pipeline industry and so are repeated here.

Regulatory Risks in General

Two features of modern rate regulation create risks that may easily go unrecognized by regulators and regulated companies alike: the differences between regulated and competitive rates, and the effect of asymmetric risks on required rates of return. Also, competition can have broader implications for regulated firms than 'mere' rate rivalry.

Competition and Regulation. Competition can bring a host of problems to regulated companies. One is the issue discussed in Chapter 4: regulation overcharges for capital assets early in their lives and undercharges for old assets, relative to competition. This difference has important implications for the risk of any regulated industry as competition becomes more important. Competition may prevent a regulated company's charging the high early rates that regulation permits on a new investment, while regulation may prevent the company's charging the high rates on old assets that competition permits.[64] The result can be assets worth 'the lower of cost or market value,' to borrow the accounting phrase.

A second problem is that regulated rates for protected monopolies typically contain rates far above cost for some customers (e.g., businesses, or long distance telephone service) in order to permit rates below cost (or not so far above cost) for other customers (e.g., residences, or local telephone service). Competition typically erodes the profit margins on

---

[64]   Also, the 'front-end load' in regulated prices distorts decisions by pipeline customers, and hence pipelines. These characteristics are economically undesirable even if growing competition were not an issue.

the far-above-cost service, a process sometimes called 'cream skimming.' If the regulated entity is not able to match the lower prices competitors charge, either because regulators forbid the necessary rate cuts or because it is less efficient, the customers providing the extra revenue shift to other suppliers, leaving the regulated entity stuck with the more-costly-to-serve customers, whose rates may not fully cover their costs.[65]

Discounting is a competitive response to such forces, but discounting by itself is at best only half the answer. Discounting permits a regulated company to keep customers that cost little to serve, but unless the company is permitted to charge full cost to previously favored customers that cost a lot to serve, the company will still be in trouble.[66]

The 'corporate culture' for regulated entities can leave them particularly vulnerable to competition, too. Typically, regulated companies evolve to treat regulators as their most important customers. After all, regulators have more to say about such companies' 'bottom line' than any other group. It has often proven difficult for regulated entities to shift gears to adopt the attitudes and flexibility needed to respond to competition for final customers.[67] This is particularly true when competition is introduced while regulation remains in force. The old skills have to remain sharp while a new set of skills is acquired. That would be a challenge for any organization.

Competition, for example, forces product diversity whether a regulated entity is interested in such diversity or not. In the long run, a 'monopoly' position at the start is no guarantee of success, or even survival, at the end; in fact, it can prove more of a hindrance than a help. Competi-

---

[65] For example, Railway Express Agency, a package delivery service operated by a consortium of railroads, went bankrupt when United Parcel Service entered as an efficient carrier. One reason was that REA retained the obligation to serve all comers, while UPS had size restrictions on the packages it would ship that permitted standardization that cut costs greatly. See Kolbe and Hodges (1989).

[66] If discounting is permitted without an upward adjustment to other rates to recover the lost revenues, the company ends up not earning its cost of capital on average.

[67] The Postal Service, for example, has steadily lost market share to its unregulated competitors despite literally decades of effort (Kolbe and Hodges 1989).

tion fosters product diversity, as companies seek out price-product feature 'niches' where they can specialize and make money. For example,
under competition, companies need to learn to pay especially close attention to costs and efficiency at the low-value end of the spectrum they
serve, while quality is far more important than cost at the high-value
end. Product differentiation is a key to competitive success. Regulated
companies, on the other hand, tend to see their product as a single-quali-
ty, homogeneous commodity, in part because rate regulation encourages
homogenization of product offerings in the name of administrative simplicity. As a result, regulated entities can fare very poorly when competition comes along (Kolbe and Hodges 1989).

All of this suggests that growing competition will bring serious risks to
the pipeline industry, if the experience of other industries is a guide.

Asymmetric Risks. The second general problem facing regulated industries (also discussed in Chapter 4) is inadequate compensation for bearing
asymmetric risks. Some risks that do not affect the cost of capital may
nonetheless affect the rate of return pipelines should be allowed. Common regulatory practice is to equate the allowed rate of return with the
best estimate of the firm's cost of capital. However, it turns out that this
practice may *not* provide investors a fair opportunity to earn the cost of
capital. In particular, if the firm faces material downside risks without
an offsetting upside opportunity, investors require risk compensation *in
addition to* an allowed rate of return equal to the cost of capital.

There are many examples of asymmetric risks in regulated industries in
recent years. Pipeline losses under take-or-pay contracts are an obvious
case in point. Pipelines earned *no* profit on the gas if it could be sold,
but were at risk of substantial losses if it could not. Disallowances of
prudently begun and prudently abandoned electric power plants are another example.

The cost of capital is a breakeven rate of return, and the average of
breaking even and incurring a loss is a loss. To break even, companies
need an *overall* rate of return, on successful and unsuccessful investments
combined, that investors expect will equal the cost of capital. That goal
cannot possibly be achieved by a policy that restricts the return on successful investments to the cost of capital, but permits returns well below

that level in other circumstances. Just as pharmaceutical companies need returns well above the cost of capital on drugs that reach the market to pay for those that do not, regulated companies need returns above the cost of capital on investments that go into the rate base if they incur large losses on other regulated activities, if they are to earn the cost of capital on average.

Pipeline Risks

We have concluded that gas pipelines are definitely *not* a low-risk industry. At the same time, it is not possible to characterize readily the level of risk currently facing the industry in any simple, dispositive way. There are two problems: first, the risks of merchant service probably differ from (and exceed, at present) the risks of pure transportation service; second, there is an implicit need for a baseline.

Transportation Risk. We have to conclude that the risks of pure pipeline transportation service are up relative to any baseline. The two sources of this greater risk are increased competition and tighter regulation. The FERC is apparently no longer inclined to permit the certification process to protect existing pipelines from new entry and new pipelines from mistaken entry. This raises the possibility of stranded investment, in which no set of rates will recover the approved cost of service.

The experience of oil pipelines, where unfettered entry has been the rule for decades, confirms that new investments can fail and old investments can be rendered unprofitable by new entry despite the physical scale economies that seem to make pipeline transportation a natural monopoly.[68] These forces have not protected oil pipelines from substantial losses, however. Oil pipeline investment has continued nonetheless for two reasons. First, it was traditionally subject to loose-handed regulation, which opened the possibility of returns above the cost of capital to offset

---

[68] Loosely speaking, the cost of a pipeline goes up with its circumference, while its throughput goes up as its cross sectional area increases. Also, the need to obtain rights of way and environmental and other permits imposes a fixed cost on any new pipeline. Thus, larger pipelines tend to be cheaper.

the risk of loss. Second, oil pipelines were sometimes viewed as invest-ments the parent companies needed to get their oil to market, and the potential value of getting the oil to market was thought worth the risk of losses imposed by competition (and perhaps more recently, tighter regu-lation).

While oil pipelines are exposed to more forms of competition than gas pipelines, that is a difference in degree, not in kind. Thus, open compe-tition for new transportation investments can be expected to expose gas pipelines to the risk of substantial losses, too.[69]

At the same time, the regulation of transportation seems to be growing tighter. Throughput targets are being set more stringently than formerly, for example. The combination of competition and tight regulation *inevi-tably* leads to asymmetric losses. Thus we conclude that transportation risks are definitely higher than they have been before.

Merchant Service and Overall Risk. The problem of a baseline becomes important in the assessment of merchant service and the overall risk of merchant and transportation service together. We have no doubt that the pipeline industry faces materially greater risks today than it did in the pre-NGPA days when the biggest problem was to find gas to satisfy the assured demand created by wellhead price regulation. The gas industry as a whole, from wellhead to burnertip, is riskier than it once was, and changes in the regulatory system have placed more of that risk on pipe-lines' shoulders.

If the baseline is the 1980s, however, we believed in early 1991 that those who argue for low risk are correct that some of the important merchant-service risks of that decade are unlikely to surface again to the same degree. New gas contracts, to the extent they exist, will be written

---

[69] Oil pipelines also lack demand charges, which tend to reduce risk, all else equal (although oil pipelines do have a partial analogue in throughput and deficiency agreements). However, discounting of demand charges has taken place for gas pipelines, a practice that is likely to grow as competition increases. And more generally, given the likelihood that pipeline firm customers would be allowed to switch service to a new pipeline if it offered better terms, demand charges in a particular year are no guarantee of capital recovery over the life of a pipeline.

with the experience of the past in mind. Also, Gas Inventory Charges, if implemented properly, could *in principle* compensate for the risks that remain. (Practice has a long way to go to implement this principle, however.)

The mega-NOPR, of course, has changed this assessment. Pipelines bore $3.4 out of $9 billion of take-or-pay losses, and preliminary estimates are that transition supply costs could be nearly $8 billion under comparability. The mega-NOPR suggests that pipelines should bear 'only' 25-50 percent of the transition costs, but *any* share of such costs is too large a share if pipelines have not been compensated for the risks. Pipelines clearly have not been.

The balance of increased transportation risk and uncertain, but perhaps equivalent, merchant risk leaves the overall level of pipeline risk perhaps roughly comparable to the 1980s, and possibly even higher. The statistical risk measures do not signal a peak in pipeline risk at present, but the most recent of those measures includes the entire 1986-1990 period, and so cannot distinguish between risk then and risk today.

Thus we have to conclude on balance that the industry today remains exposed to substantial risks overall. The most important include uncertain regulatory rules and growing competition (including comparability), as well as 'wild cards' such as environmental cleanup liabilities.

Regulated industries generally learned during the 1980s that regulation does not protect against massive losses, and pipelines are certainly no exception to that discovery. Nor do traditional regulatory procedures offer fair compensation for the risk of such losses. For example, traditional regulatory procedures contained nothing like a GIC, or indeed, any mechanism at all,[70] as compensation for the risk of take-or-pay losses, so pipeline allowed rates of return during the period of take-or-pay expo-

---

[70] In principle, allowed rates of return could have been set above the cost of capital as compensation for take-or-pay risks, although that would be a difficult approach to follow because the magnitude of the take-or-pay risk had little or nothing to do with the magnitude of the rate base on which a return is allowed. In practice, however, allowed rates of return displayed no such premium, nor have we found any evidence whatsoever that the FERC intended the rates of return it allowed to include a premium above the cost of capital.

sure were surely inadequate. (To use allowed rates of return from the past as a benchmark for returns today is simply fallacious.)

Nor do claims that explicit changes in regulatory policy and the nature of the pipeline business now make pipeline risk low withstand scrutiny. That those changes might have done so if everything else had remained equal is irrelevant, because everything else did *not* remain equal. Nor is there any reasonable assurance that GICs will be set at compensatory levels in practice, and the concern that pipelines might make a 'profit' on GICs, an absolute necessity if GICs serve an insurance function and the insured risk does not occur, is reason to believe they will not. And comparability can only make pipeline risks worse, at least on average, on existing gas contracts.

A possible response to such problems is a loosening of the regulatory constraint, to give pipelines some upside potential to balance the downside risk. Yet the picture we perceive is one of *tighter* regulatory constraint, through the use of more stringent throughput forecasts to set rates, for example. Just as importantly, rule changes have been sufficiently frequent that pipelines must fear that even 'safe' investments may become 'risky' on short notice. The mega-NOPR on sales comparability clearly has this potential, for example, as discussed in more detail below.

Growing competition is another major source of risk. Importantly, proposals now exist to increase competition still further for both the transportation and sales parts of pipelines' business. Even if these proposals are not implemented, pressure for increased competition from both regulators and the marketplace is likely to persist. Competitive returns will tend to be more sensitive to the business cycle than regulated returns, all else equal, which will tend to increase risk and the cost of capital.[71] Also, competition ordinarily creates the potential for large losses; for example, a competitor may usurp a company's position with long-time customers. In principle, if competition is the only constraint, it provides an offsetting opportunity for large gains -- the business one pipeline loses can be another pipeline's gain. Regulated companies often

---

[71]   All else may not be equal under full deregulation, because full deregulation permits companies to abandon unprofitable services in bad times. However, full deregulation does not seem to be contemplated for gas pipelines.

seem to have difficulty seizing such opportunities, however. The skills needed to satisfy regulatory requirements are not necessarily the skills needed to respond quickly to competitive threats and opportunities.[72]

The major problem that competition creates for pipelines, however, is that competition is *not* their only constraint. Pipelines increasingly face two masters, the market and rate regulation. Most of American industry faces one or the other, but not both. These masters discipline rates differently, creating the potential for large losses without the offsetting potential for large gains.

We noted above that product differentiation is a key to competitive success. The reason is simple: product differentiation creates a form of market power, permitting competitive firms to earn more than the cost of capital until competitors offer products similar enough to erode those profits. The relentless drive for differentiated products is one of the key benefits of competition, since it provides new goods and services.

Firms facing regulation and competition together, however, require services with some degree of market power for another reason. Competition can prevent regulated firms from offering service at rates that regulation will permit. The widespread discounting gas pipelines do today is ample evidence of this fact. If pipelines are to have a fair opportunity to earn the cost of capital despite competition, there *must* be services where they can earn high markups. If *every* service the regulated company provides is subject to competition, then as discussed in the next chapter, either regulation must change the way it sets rates or the industry will inevitably decline.

For example, the ability to offer firm sales service on better terms than unregulated competitors is a differentiated product of the type that regulated companies *must* have when exposed to both competition and traditional regulation, if they are to continue to invest in the industry. The mega-NOPR on comparability suggests that the FERC is determined to

---

[72] However, to the extent that growing competition for sales is forcing pipelines to hone those skills, they are likely to be applied to transportation as well. The result will ultimately be increased competition in both areas, even without further rule changes.

do away with such products. As discussed in more detail in Chapter 6, the Commission cannot do so across the board unless it is willing either: (1) to make major changes in the way it regulates the pipeline industry, or (2) to see the industry decline.

Thus, we find we side with the pipelines' view on their risks on balance. We have concluded that the pipeline industry cannot fairly be characterized as 'low risk' overall, and in some dimensions, particularly exposure to regulatory rule changes and to a mix of regulation and competition, it is definitely high-risk. The view that the industry's troubles are in the past and that regulators can be counted on to fix any remaining problems is simply not borne out by the evidence.

But one argument remains unanswered in the response of our 'regulators,' and that is the most fundamental of all: if pipelines are not being adequately compensated for risk, why do they continue to make new investments? That argument is the springboard for what we have come to believe are the pivotal issues facing the industry and its regulators. These are the topic of our next chapter.

Chapter 9
Two Fundamental Questions

Homeowners in Chicago, seduced by falling electric rates, are abandoning gas at a four-per-week pace, while Chicago's electric utility, Commonwealth Edison Co., is signing up some 70 new residential customers a week. . . .

As for gas producers, their number has plunged 60% since 1985. They are going out of business at the rate of one a week, and the pace is expected to quicken. Many gas producers are writing down assets, renegotiating loans and selling off properties they can't afford to explore and drill.

All this will eventually mean less drilling, shorter supplies and still-higher retail prices. Already, drilling for gas has all but stopped in the U.S. (Johnson 1992, A1)

Chapter 8 concluded that pipelines' transportation service is definitely growing in risk and that pipelines are a risky industry. The level of take-or-pay risk was down from the level of a few years ago, but a transition to sales comparability reopens this issue. This does *not* imply that pipeline rates of return can be left at, let alone reduced from, the level of the mid-1980s, because those rates of return clearly were *not* compensatory for the risks the industry faced at the time. Nor do the regulatory and business changes the industry is going through imply pipelines are now a low risk industry. Very substantial risks remain, particularly from uncertain regulatory rules and growing competition, certainly more than existed in the pre-NGPA days.

Chapter 8, however, left one issue open in its imaginary conversation between pipelines and 'regulators': why pipelines continue to invest if their risk compensation is inadequate. This issue points the way to two truly fundamental questions that face the industry and its regulators today:

- Given the level of risk to which the industry is now exposed, whatever it is, do current regula-

tory and legal procedures permit investors to
expect fair compensation for that risk? In other
words, is the current system *sustainable*?

- Regardless of whether the current system is
  sustainable, are there aspects of the system that
  should be changed? In other words, is the current system *desirable*?

To set the stage for these questions, we first must address the final issue
from Chapter 6.

"WHEN YOU STOP MAKING NEW INVESTMENTS, THEN WE'LL START TO WORRY"

Our regulators' final response to pipeline claims of high risk now must
be faced head on: if the problem of asymmetric risk is as bad as we say
it is, why are pipelines making new investments? There are at least four
possible answers:

1.   The problem is not as bad as we claim, even in
      principle;

2.   The problem is as bad as we claim *in principle*,
      but in practice no asymmetric risks are impor-
      tant to the industry at the moment;

3.   The problem is as bad as we claim, but pipeline
      management is only gradually becoming con-
      vinced it will really apply to transportation
      investments as well as gas sales; and

4.   The problem is as bad as we claim, but pipeline
      management deals with it by making only invest-
      ments that they believe have little or no asym-
      metric risk.

A combination of the third and fourth answers is the correct explanation.
Here's why.

The electric utility industry at present is evidence that the problem is as bad as we claim in principle. Our experience with that industry, which includes conversations with many utility executives, regulators, and industry advisors, and which we do not believe is atypical, tells us that utilities are extremely reluctant to build new capacity. Once demand is virtually assured, and if third parties cannot be found to satisfy that demand, and if the capital required is modest, utilities are willing to build, but not until then. That is the logical and economically rational consequence of the 'heads I break even, tails I lose' compensation system that evolved in the 1980s: invest only as a last resort, when you are sure you cannot lose.[1]

Gas pipelines, like electric utilities, learned in the 1980s that the limits regulation imposes on profits are not matched by equal limits on losses. Their losses came in the form of gas purchase costs rather than disallowed investments. But for pipelines, too, a substantial adverse change in the underlying economics of a regulated business turned out to cost regulated shareholders large amounts of money, even though equivalent gains would not have been available from a substantial favorable change. Management forgets such lessons only at the expense of the shareholders who hire them, and modern shareholders seem more than willing to sue to remind them, if necessary.[2]

However, this does *not* mean that no investment will take place; it means that only *safe* investment will take place. Here 'safe' does not mean 'risk-free'; it means free of *asymmetric* risk to investors. The invest-

---

[1]   One could imagine regulatory penalties for *failure* to invest a few years down the road. If so, the optimal utility strategy is to get out of the generation business entirely. Moreover, as a simple practical matter, penalties for failure to build are not as great a risk. First, a percentage point or two off the allowed rate of return on equity for failing to invest is likely to be a much smaller penalty than the loss of a major portion of the cost of a large new power plant. Second, if regulators want the utility to have access to the capital needed to catch up with demand, they cannot penalize the utility too severely.

[2]   Thus, a fifth answer to why pipelines are making new investments is that pipeline management is doing so irrationally, against the shareholders' interests. We have no evidence that favors such an explanation. Moreover, even if it were true, this is not an answer on which policy makers can rely, because such management cannot survive.

ments chosen will be subject to the normal risks of the pipeline business, but managers will not knowingly make investments with any risk of substantial downside loss, because the current regulatory scheme tries to insure there will be no corresponding upside reward.

Recall, for example, regulators' concern that pipelines might make a profit on Gas Inventory Charges! As noted in Chapter 8, they had better be allowed to do so if no contract loss occurs, else GICs are *guaranteed* to be noncompensatory. Also, we understand that throughput assumptions on new pipeline investments are becoming ever more stringent. As also noted in Chapter 8, the closer the assumed throughput approaches 100 percent of capacity, the greater the potential asymmetric risk. The more stringent the throughput assumptions become, then, the more restricted are the investments that pipelines will make.

Nor are regulatory actions the only asymmetric risks pipelines currently face. As discussed in Chapter 4, the combination of competition and regulation inherently creates asymmetric risks. Figure 9-1 recreates Figure 4-4, which shows how competition can prevent new pipelines from earning the rate of return regulation would permit, while regulation can prevent old pipelines from earning the rate of return competition would permit.[3] The result is that investments can be worth 'the lower of cost or market,' indicated by the bold line in Figure 9-1. This means some pipeline investments that would make society better off will not be made, because they would compete with an old, largely written off pipeline that provides service less efficiently but at a price well below the competitive market price.

Thus, important asymmetric risks exist, risks that will definitely influence pipeline behavior. However, investments without material known asymmetric risks will also exist, and pipelines can be expected to go ahead and make such investments when possible. The existence of continuing pipeline investment is consistent with asymmetric risks' existence and with their having a material influence on pipelines' choice of investments.

---

[3]   Recall that nominal capital charges are in dollars of declining purchasing power. In dollars of constant purchasing power, the 'competition' line would be flat and the 'traditional regulation' line would fall much more rapidly than in Figure 9-1.

Figure 9-1

**Capital Recovery May Be Impossible for Firms Facing
Both Regulation & Competition**

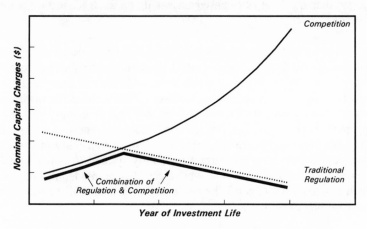

Having thus responded to the final comment from our 'regulators,' we now have the background to address the two fundamental issues.

IS THE CURRENT SYSTEM SUSTAINABLE?

As used here, a 'sustainable' regulatory/pipeline system is one that meets demand and compensates investors adequately for the level of risk they bear. A system that does not meet the latter criterion will fail to attract new investment and eventually will decay.

The current system is generally sustainable in the short run, providing material adverse rule changes such as large transition losses under comparability are avoided, because pipelines have substantial sunk investment that will not vanish overnight. Pipeline management can decide whether or not to make new investments. Pipelines with a largely written off existing rate base who can roll the cost of the new investment in with the old rate base may even be able to make quite substantial new investments.

It is doubtful that the system is sustainable in the long run, however. Minimum-risk investment strategies tend to lead to capacity shortages and poor service, which in turn can reduce demand and further retard investment. Concern over gas availability is already a factor in interfuel competition, and underinvestment in gas pipelines can only make that problem worse. The tighter the regulatory constraint, the greater the danger that the long-run consequence will be an inadequate pipeline network.

In industries with long-lived investments, it can take decades for such problems to manifest themselves, and still more decades for their correction. Consider how long the nation's railroads took to decay and how slow the recovery has been. Another example is the nation's roads and bridges, where the costs but not the benefits of timely maintenance were apparent to the responsible governmental agencies. These examples are cause for serious concern about the long-run sustainability of the gas pipeline network under the current regulatory system. It is in no one's interest to wait for the pipeline equivalent of standing derailment (i.e., when a locomotive falls off the track while standing still) to recognize that a problem exists.

IS THE CURRENT SYSTEM DESIRABLE?

We have no uncertainty about whether the current system is desirable. The answer is a flat 'no.' There are two main problems: economically inefficient investment incentives and economically inefficient pricing.

Inefficient Investment Incentives

In practice, much of the current regulatory system for pipelines aims at ensuring pipelines do not expect to earn more than the cost of capital. Yet it provides no similar assurance that pipelines do not expect to earn less than the cost of capital. Thus, if a pipeline identifies an investment that may turn out to provide exceptionally low cost service, but has a small risk of failure, the pipeline undertakes it only to the disadvantage of its shareholders. If it does turn out well, regulators will pass the benefits through to customers in the form of low tariffs in the next rate case. If it turns out poorly, some or all of the costs go to the investors.

Even if customers would gladly approve the investment and willingly bear the small cost of failure, pipelines have the incentive to avoid it, because they share the risks but not the rewards.[4]

This cannot possibly be optimal from society's viewpoint. Some risks are worth taking, because the expected value of the outcome is very positive, even given the risk. A system that gives the rewards to one party and the risks to another discourages the taking of such socially desirable risks.

For example, the Commission's recent decision in *Transwestern*,[5] discussed in Chapter 5, left the pipeline at risk if the capacity of the system turned out to exceed demand. We are unfamiliar with the details of that case, and so cannot say whether that decision was economically optimal in the circumstances. But in general, there are likely to be cases where the total cost to society would be lower if the pipeline were somewhat oversized initially, in anticipation of growth in demand, rather than to have to add pumping capacity or loop the existing line each time demand grows. Yet if pipelines were at risk for unused capacity but customers got all the benefits from cheaper incremental capacity, pipelines would penalize their shareholders if they risked building in excess of identified demand.

The analogy to the contingency fee discussion in the concurring opinion by Judge Williams, raised briefly in Chapter 6, is instructive.[6] Suppose a plaintiff's attorney is deciding whether to take a case, and suppose the usual, noncontingency fee (the 'lodestar' fee, in the vernacular of these cases) would be $10,000 for a case requiring this amount of effort. That

---

[4] Of course, regulatory lag can give pipelines a modest share of the rewards, but in the long run under an asymmetric system the benefits of exceptionally fortunate investments go to customers while the costs of exceptionally unfortunate ones are shared by investors.

[5] *Transwestern Pipeline Company*, Docket CP90-2294, issued January 17, 1991.

[6] *King* v. *Palmer*, 906 F.2d 762, 769-70 (D.C. Cir. 1990) (concurring opinion). Again, Judge Williams (1991, 159, 161) explicitly notes the analogy to the regulatory allowed rate of return in his review of a paper by two of the authors of this book.

is, if the attorney turns down this case, he or she can expect to spend those hours on a noncontingency fee case worth $10,000 instead. If the fee depends on success, the attorney will take the case for a $10,000 fee only if success is assured. Since there are few cases where success is assured, few plaintiffs' cases would be taken.

Recognizing this, the law permits a 'contingency enhancement' to the 'lodestar fee.'[7] If the contingency enhancement is set at one-half of the lodestar fee, plaintiffs' attorneys will break even on average if they take cases with a two-thirds probability of success. Thus, if they take 15 cases like this over a period of time and win 10 of them, and if in each winning case they get $15,000[8] and nothing if they lose, they receive $150,000 from the contingency work. For noncontingency work, they would have received $10,000 for each of the 15 cases, or the same $150,000. If the contingency enhancement is set at 100 percent of the lodestar fee, the usual practice according to Judge Williams's opinion,[9] plaintiffs' attorneys will break even if they take cases with a 50 percent chance of victory.[10]

An asymmetric system for regulated companies is equivalent to attorneys' taking contingent cases but getting only the lodestar fee if they win. The rational response is to accept only 'sure things.'[11] Or, in terms of one of the earlier examples, an asymmetric system is equivalent to drilling for oil and earning only your cost of capital on gushers. Clearly, if you are to earn the cost of capital on average and if one outcome is a dry hole at a substantial loss, in the gusher outcome the return has to be well above the cost of capital.

---

[7]   The sum of the contingency enhancement and the lodestar fee is called the 'contingency fee.'

[8]   That is, a $10,000 lodestar fee and a $5,000 contingency enhancement.

[9]   *King* v. *Palmer.*

[10]   Winning 10 of 20 cases with a $20,000 contingency fee is the same as working on all 20 for a lodestar fee of $10,000.

[11]   Here, 'sure thing' does not mean without risk at all, but rather without downside risk that is not balanced by upside potential.

Society has decided that some level of contingent lawsuits are desirable, so society authorizes contingency enhancements. Similarly, society finds oil, pharmaceuticals, and software packages desirable, so society does not tax away all profits above the cost of capital on successes in these industries. If it did, no one would invest in them because the average of a loss on failures and the cost of capital on successes is less than the cost of capital.

Society trusts market forces to equilibrate the risk-reward tradeoff for goods like oil, pharmaceuticals, and software.[12] As Judge Williams points out, society instead sets the tradeoff for contingent litigation by the court-approved contingency enhancement markup.[13] As Judge Williams (1991) also points out, society sets the tradeoff for regulated investments in an asymmetric system by the degree to which it permits regulated companies to earn more than the cost of capital on successful investments. But under the current system, that decision is to encourage pipelines to accept *no risk of failure at all*.

As noted above, that cannot possibly be the economically optimal decision. As investors avoid asymmetric risk, a new kind of asymmetric risk arises for customers. Sudden surges in demand may not be met, because undercapacity is safer than overcapacity, which risks disallowances for investors. Also, investments that are expected to improve service substantially, but which have a small but noticeable chance of failure, will not be undertaken. Ultimately, gas is likely to be less available and less reliably available than alternative fuels that do not face such a constraint.

Inefficient Pricing

The second major problem with the current system is inefficient pricing. As discussed in Chapter 4, the competitive price is independent of the vintage of the assets owned by any competitor, while prices of companies under Original Cost regulation depend heavily on the vintage of the assets employed. (Colloquially, the farmer gets the same price for toma-

---

[12] Albeit with some special tax provisions to shift the tradeoff in some areas.

[13] *King* v. *Palmer*.

toes whether the farm was bought in 1891 or 1991, but pays a very different price for power depending on whether the power plant was built in 1971 or 1991.) This creates two problems for the pipeline industry and its customers.

The first is that Original Cost regulation encourages overuse of old assets and discourages use of new assets. The result is likely to be life extensions where asset replacements would be more efficient, and routing gas over old systems even when routing over a new system would be more efficient.[14] The result is also to overprice gas shipped on new systems and underprice gas shipped on old systems, which distorts the choice between gas and alternative fuels. That in turn can lead to inefficient investment choices by end users, choices which, once made, distort consumption for years afterwards.

The second problem is that mentioned above: Original Cost regulation and actual competition can combine to make economically sensible investments, from society's viewpoint, into economic suicide from the investor's viewpoint. New investments will be made only in areas that lack effective competition or by companies that can roll in the high cost of the new facility with old, low-cost facilities. These additional constraints on what investments can be made[15] are sure to lead to a socially less desirable mix of investments than an unconstrained system, even if there were no other incentive problems at all.

THE CONSEQUENCES OF INACTION

The kinds of effects discussed in this chapter are likely to take a long time to produce large, undeniable economic consequences. Successive regulators can let the problem slide with only modest incremental damage, as successive generations of administrators responsible for maintenance of the nation's roads and bridges have. Similarly, for many years,

---

[14]   In this context, 'efficient' means lowest *economic* cost. Economic cost values assets at their competitive market value, not at historical cost.

[15]   That is, the requirement that there be no effective competition or that the price be able to be rolled in with an existing system.

the largest single item of cash flow on railroads' book reportedly was an item called 'deferred maintenance.' Letting tomorrow's taxpayers (or shippers) pay for the problem is an easy and in many ways a safe course. And given the workload that the day-to-day responsibilities of a regulator's or legislator's job entails, it would be more than understandable if the first reaction to these problems were to want to leave them to successors.

The issue in many ways is what the policy makers see their job to be. In the day-to-day conduct of rate regulation, the task in recent years has often seemed to become to ensure regulated investors are not overcompensated and that their customers do not pay too much. As a response to some situations, such as the overcapacity of the electric industry in much of the 1980s, this is again a more than understandable focus. But as economists, we would argue that customers of regulated industries are best served by the *most efficient* system, not by the lowest-cost *no-risk* system, which is sure to cost more in the long run. Thus, achievement of the most efficient system will require the bearing of some risks.

Investors are the right party to bear risks, not final customers. Investors do so voluntarily, with money that does not come out of the food budget. But to induce investors to bear risks, the investment must offer an upside opportunity commensurate with its downside peril. The potential for downside peril shifted during the 1980s, exposing investors in regulated companies to more risk than at any time since the Great Depression. Some of that risk can be expected to pass as the economic conditions that created it pass. But some will remain, if regulation or regulation and competition together create an asymmetric system of payoffs.

Policy makers can ignore this fact only at substantial long run harm to the customers of the industries they regulate. That harm may be a long time in coming in an industry with long-lived investments, but if it is allowed to happen, the remedy will be a long time coming, too. Preventive maintenance on the gas pipeline regulatory system is clearly in order now.

There is widespread belief that natural gas has environmental advantages over oil and coal and national security advantages over oil. If these advantages are to be realized in actual switches to natural gas, a solid gas pipeline infrastructure must remain in place. The authors therefore believe the applicable policy makers should address the problems raised here as soon as possible.

Appendix A
Risks That Affect the Cost of Capital

Chapter 4 summarized the difference between diversifiable and nondiversifiable risk. This appendix examines this difference in more detail. The appendix also describes an important subdivision of nondiversifiable risks, *business* and *financial* risks.

RISK AND SECURITY PRICES

Much of finance theory focuses on the forces that determine security prices, and relative risk is a major determinant. Security prices are driven by the cash flows investors expect and by the rate at which investors discount future, uncertain cash flows. That rate, the cost of capital, depends in turn on the relative risk of those cash flows.

The pipeline industry risks described in this report all have the potential to affect a pipeline's cash flows. Price cutting by a business competitor cuts a pipeline's earnings, a major component of cash flow. A regulatory penalty may do the same. And an operational problem, such as an undetected mismatch between the gas taken and the gas delivered on a particular contract, may leave the pipeline to make up the difference out of its own pocket.

Since these risks affect cash flows, they affect realized rates of return. There are many rates of return that might be of interest: rate of return on book equity, rate of return on rate base, rate of return on debt, and rate of return on a stock or bond. Since finance theory focuses on capital markets, it is the rate of return on securities that is of immediate interest. If a piece of bad news cuts current or expected cash flows, whether its source is business, regulatory, or operational risk, security prices will be lower than they would have been without the bad news. Conversely, good news lifts security prices, all else equal. Thus risk, whatever its source, translates into volatility of security prices. If we seek to determine the rate of return investors require, a natural first measure of the risk of a company therefore is the volatility of its securities or of the rates of return on those securities.

Total Volatility as a Risk Measure

Suppose for now that the company is financed entirely by common equity. Then the volatility of its shares measures the volatility of the present value of the company's expected future cash flows. The total volatility of those shares therefore is a natural first measure of a company's risk.

If the possible rates of return shareholders might earn are thought of as lying on a bell-shaped curve, the volatility of the shares depends on how wide the bell-shaped curve is. For example, the two curves in Figure A-1 both represent distributions of possible rates of return for a common stock, and each has the same average value -- 15 percent. But one distribution is wide and flat, while the other is narrow and pointed. The stock with the wide and flat distribution stands a much greater chance of an extreme rate of return, one very far from the expected value of 15 percent. In other words, it is more volatile, and hence it is riskier.[1]

Thus high volatility for a stock's rate of return indicates a stock is very risky. But while high volatility suggests investors will demand a high expected rate of return, it does not *necessarily* mean they will. Volatility is not the whole story for risk measurement.

Diversification as a Way to Reduce Risk

Chapter 4 used the example of the gambling casino to illustrate diversification. The undiversified gambler had considerable risk from roulette, but the casino was diversified against this risk. Yet the casino still faced considerable risk due to swings in the economy and the resulting impact on the amount bet. Such non-

---

[1]   This greater risk can be quantified by calculating the *standard deviation* of the two distributions, a common statistical measure. The standard deviation of the wide distribution is larger than that of the narrow distribution.

Figure A-1
Possible Rates of Return

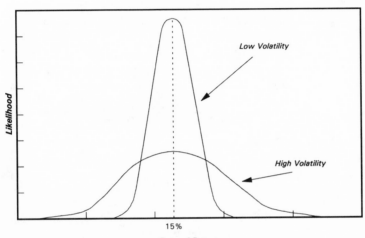

diversifiable risks were the ones of real concern to the casino's investors.

Chapter 4 also illustrated that the combining of stocks into portfolios eliminated much of the risk of individual stocks. Figure A-2 illustrates this point.[2] Again, by the time ten to fifteen randomly selected stocks are added, a material portion of the risk of the individual stocks is eliminated.

Where the portfolio's volatility bottoms out depends on the degree of correlation among the stocks in the portfolio. The sometimes dramatic movements of the stock market as a whole indicate that even a portfolio of *all* stocks is still risky. As noted in Chapter 4, the risk that cannot be eliminated even in large portfolios is known as *nondiversifiable* or *systematic* risk. This is the risk

---

[2]   Recall that standard deviation is just a measure of the amount of volatility. The lower the standard deviation, the less volatile the portfolio's returns.

Figure A-2
Impact of Number of Stocks on a Portfolio's Risk

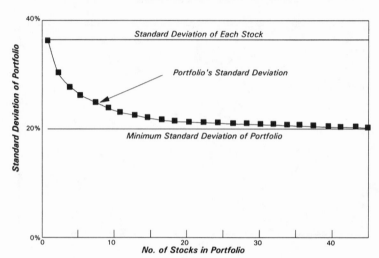

that exists even for well-diversified investors, and therefore the cost of capital will include compensation for nondiversifiable risk if investors are risk-averse (as the evidence indicates they are).

The basic measure of nondiversifiable risk is known as *beta*. Beta measures the degree to which a stock tends to move up or down when the market as a whole moves up or down.   Stocks with betas equal to 1.0 have average risk by definition, tending to rise or fall 10 percent when the market rises or falls 10 percent.   Stocks with betas above 1.0 exaggerate the swings in the market: stocks with betas of 2.0 tend to fall 20 percent when the market falls 10 percent, for example.   Stocks with betas below 1.0 are less volatile than the market.   A stock with a beta of 0.5 will tend to rise 5 percent when the market rises 10 percent.

Beta is a widely accepted risk measure, probably the best single risk measure for a stock.   Many services report betas, including Merrill Lynch's quarterly *Security Risk Evaluation* and the *Value*

Line Investment Survey. Betas are not always calculated the same way, and therefore must be used with a degree of caution, but the basic point that a high beta indicates a risky stock is widely accepted by both financial theorists and investment professionals. (Various theories of just how the cost of capital relates to beta, such as the Capital Asset Pricing Model,[3] remain controversial in finance; but the importance of beta as a risk measure does not.) Thus the importance of nondiversifiable risk to investors is widely accepted, too. The large, active, well-diversified investors in the market eliminate compensation for diversifiable risk, but even they must bear nondiversifiable risk.

BUSINESS VERSUS FINANCIAL RISK

In the businessperson's lexicon, 'business risk' often means risks driven by general business conditions, the responses of competitors, and the like. In the financial economist's lexicon, 'business risk' appears as a subcategory of nondiversifiable risk. *Business risk* in this sense is the (nondiversifiable) risk investors would bear if the company were financed entirely by common equity. *Financial risk* is the extra risk shareholders bear when the company issues debt to finance part of the business. The cost of equity capital increases with either source of risk.

The source of business risk is the uncertainty in sales and operating costs. Thus it includes the day-to-day impact of the regulatory rules under which a regulated company operates. It may include operating risks as well, if the level of risk is correlated with the economy. (For example, one could imagine that customers or producers might be more likely to make 'mistakes' in deliveries or takes in especially good or bad economic times.)

---

[3]   The Capital Asset Pricing Model says the cost of capital for a stock equals the sum of the risk-free interest rate plus the product of the stock's beta times the 'market risk premium,' the risk premium investors require for investments of average risk. See, for example, Brealey and Myers (1988, Chapters 7 to 9).

The source of financial risk is the prior claim that bondholders have on the resulting earnings, and ultimately on the company's assets.   This claim magnifies the impact of the basic business risk on the shareholders, and so increases the overall nondiversifiable risk they face.

An example of financial risk is given in Table A-1, which ignores taxes for simplicity.   Panel A shows alternative balance sheets for the same company.   Both have $1,000 in assets, one consisting entirely of equity and the other with one-half debt financing.   Panel B is an abbreviated income statement.   The operating earnings are expected to be $150.   For the purposes of this illustration, assume that the risk of these operating earnings can be adequately described by specifying a high and low possible value, as indicated in the table.   Assume also that the implied 15 percent rate of return on total assets equals the cost of capital that reflects the business risk of this company.   If there is no debt, equityholders bear only this business risk and receive the entire operating earnings of the company.

If half of the company's assets are financed by debt at a 10 percent interest rate, however, bondholders have a prior claim on the first $50 of operating earnings.   The net income to equityholders is reduced by $50 in each case, but equityholders still bear the entire variability in the firm's operating earnings.   As can be seen in Panel C, instead of expecting 15 percent plus or minus 10 percent, equityholders now expect 20 percent plus or minus 20 percent.   That is, equityholders will be better off on average but may end up with nothing at all, even though the business risk of the company as a whole is unchanged.   The extra five percentage points (20 percent - 15 percent) in their expected rate of return is fair compensation for the additional financial risk that equityholders bear when the company is 50 percent debt-financed.

The cost of equity has gone up, but as shown in Panel D, the overall cost of capital remains the same.   This clearly makes sense, because the overall business risk of the firm has not been changed by the addition of debt.

TABLE A-1		
EXAMPLE OF FINANCIAL RISK		
	Capital Structure	
	All Equity	*50 Percent Debt*

Panel A: Balance Sheet

Equity	$1,000	$500
Debt	0	500

Panel B: Income Statement

Operating Earnings		
High	$250	$250
Expected	150	150
Low	50	50
Interest Expense	0	
		50
Net Income		
High	250	
Expected	150	200
Low	50	100
		0

Panel C: Rate of Return of Equity

High	25%	40%
Expected	15%	20%
Low	5%	0%

Panel D: Overall Rate of Return

	Cost	Share	Component	Cost	Share	Component
Equity	15%	100%	15.0%	20%	50%	10.0%
Debt	---	0%	0.0%	10%	50%	5.0%
TOTAL			15.0%			15.0%

Financial risk exists for most stocks. Therefore, the average-risk beta of 1.0 reflects a degree of financial as well as business risk. Possible differences in the degree of leverage and financial risk need to be kept in mind when comparing the relative risks of individual companies and industries.

SUMMARY

The first question a financial economist asks about a particular risk is whether it is likely to be diversifiable or nondiversifiable. If it is nondiversifiable, it will matter to investors and hence will increase the firm's cost of capital (although the precise relationship between the degree of non-diversifiable risk and the cost of capital remains uncertain). If it is diversifiable, it will not affect the cost of capital (although it may still affect investors' required returns, as discussed in the last part of Chapter 4). Use of debt magnifies the risk equityholders face, because payments to bondholders are fixed (absent default), and levered equityholders must bear the entire business risk of the firm on a smaller equity base.

Appendix B
Summary of Recent History of the Interstate Natural Gas Pipeline Industry[1]

> Regulation of natural gas markets in the United States historically has been memorable primarily for its poor performance. During the past half-century, natural gas markets have been the scene of acute shortages, prices perpetually out of step with market conditions, rancorous litigation, and intense political infighting (Kalt and Schuller 1987b, 1).

BASIC STRUCTURE OF THE GAS INDUSTRY

The chief actors in the natural gas industry are producers, interstate natural gas pipelines, local distribution companies (distributors), and the ultimate customer at the burnertip. Although individual firms are vertically integrated in all three phases in individual situations, a typical pattern might well involve firms operating at only one level in the vertical chain. At the wellhead would be the gas production firm (possibly a pipeline affiliate), which traditionally had a contractual relationship to sell the gas to the next stage, an interstate gas pipeline. The pipeline then transported the gas to the distributor, which provided a service of interconnections with the end user at the burnertip. Sometimes, one pipeline might sell the gas to another pipeline in between, or the

---

[1] This brief appendix is designed to give only sufficient overview of the recent past of the interstate pipeline industry to give context to an assessment of the magnitude of future risk relative to past risk. Those interested in greater detail should consult Kalt and Schuller (1987a); Pierce (1988); and Carpenter, Jacoby, and Wright (1987). The present appendix is designed to provide an objective history of the most significant developments while avoiding any assessment of blame for these unfortunate events. Thus we have relied to the extent possible on the findings of fact in recent court reviews of decisions by the Federal Energy Regulatory Commission (FERC), in particular those of Judge Stephen F. Williams in *Associated Gas Distributors* v. *FERC* (AGD-I), 824 F.2d 981 (D.C. 1987). Of course, the authors of this study have made their own observations of what these events mean in terms of shifting risk incidence. Finally, this appendix has not been updated to discuss the July 1991 mega-NOPR on comparability, which is covered to the extent possible in the main text.

second pipeline might sell pure transportation service to the first. Also, in some cases a large industrial company might be served directly by a pipeline.

Until recently, distributors purchased their supplies of gas from interstate pipeline companies and passed these costs along, generally automatically and without mark-up, to their retail customers, using 'purchased gas adjustment' (PGA) rate charges. Historically, distributors were ordinarily 'full requirements' customers of the interstate pipeline. The price the pipeline charged was regulated for reasonableness and prudence by the FERC. The interstate pipeline recovered gas acquisition costs under a purchased gas adjustment mechanism similar to that of the distributors. Distributors often were interconnected with only one interstate pipeline and often would find construction of alternative facilities very expensive or even prohibitive. They were also bound by long-term contracts with their pipeline suppliers that typically specified 'demand charges' -- fixed fees to pay for capacity made available, which are independent of actual sales -- and until recently, 'minimum bills' or guaranteed purchases.

RISK ALLOCATION BEFORE ORDER NO. 380

Much of the job of regulators over the last four decades has been to assign risks of gas price volatility and supply availability along the vertical chain in Figure B-1 (Stalon 1990). Until recently, pipelines primarily acted as merchants and transporters of gas they owned. They purchased gas from producers and other pipelines and offered it for resale in interstate markets. The transportation function was, for most gas and most customers, part of the 'bundle' that was purchased along with the commodity. A relatively small amount of gas belonged to third parties, who paid a pure transportation charge for moving their gas through interstate gas pipelines.

Figure B-1

Natural Gas Markets

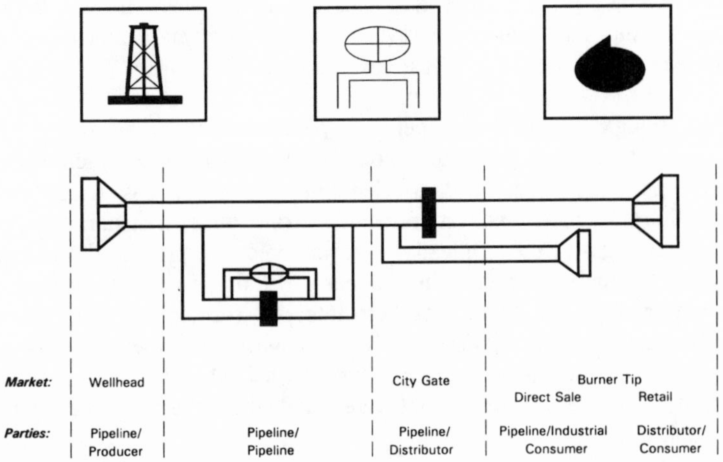

Market:	Wellhead		City Gate	Burner Tip	
				Direct Sale	Retail
Parties:	Pipeline/	Pipeline/	Pipeline/	Pipeline/Industrial	Distributor/
	Producer	Pipeline	Distributor	Consumer	Consumer

Source: Hall, "Getting Regulation from 'Here' to 'There'"

The price of gas in interstate commerce was regulated at the wellhead by the FERC or its predecessor, the Federal Power Commission.[2] For many years, that price remained well below that of competing fuels on an energy equivalent or Btu basis. As merchants, pipelines acted primarily to acquire gas for their wholesale customers and to match capacity to their customers' projected total requirements. Since interstate gas was priced below its market value, the main problem facing pipelines was to obtain adequate supply. Thus much of the industry's regulatory effort went to persuade FERC and its predecessor to encourage exploration and production and to encourage producers to dedicate reserves to the interstate market.

---

[2] This regulatory responsibility was imposed in *Phillips Petroleum Co.* v. *Wisconsin*, 347 U.S. 672 (1954). Experts have attributed much of the regulatory failure in this industry to this ill-fated decision and later discrepancies between regulated and competitive prices for gas. See, for example, Pierce (1988) and Coucil on Economic Regulation (1989).

In 1972, the Commission enacted its purchased gas adjustment
mechanisms, which automatically passed through the cost of gas
to customers. The Commission in theory could deny pipelines
the recovery of gas costs that were found to be imprudently
incurred. In practice, the fact that gas prices were heavily regulated meant that prudence reviews were rare and almost never
resulted in the denial of gas cost recovery.

Most nongas costs of pipeline operations were recovered in
charges to the customers to whom pipelines sold gas (often called
'tariff gas' or 'sales gas' to distinguish it from gas transported
for others). The modest revenues from pure transportation of
third-party gas were typically just credited to pipelines' cost of
service. Pipelines could, in the early 1980s, choose to transport
or not transport as it suited their interests (subject, of course, to
FERC approval). Consequently, transportation was a tool to
maintain or increase throughput which helped pipelines meet or
exceed their target throughput rates and thus their allowed rates
of return.

To obtain a commitment from a producer, the pipeline ordinarily
needed to sign a long-term gas purchase contract, thereby incurring risk. However, the pipeline typically hedged this risk in
one of two ways: either it was the only possible supplier for
a particular customer, or it had minimum commodity bills (promises that a customer would pay a certain minimum annual amount
for gas whether that customer took delivery of the gas or not)
to cover the pipeline's contract with the producer.

The allocation of risk that emerged from this system was two-
fold:[3]

> 1.   On the supply side, a pipeline could not
>   get a certificate of public convenience and
>   necessity unless it could demonstrate that it
>   had large, long-term reserves of gas under

---

[3]   For further discussion, see Stalon (1987) and Legato (1987).

contract to meet the needs of pipeline customers;

2.   On the demand side, regulators protected pipelines against competition by:

- limiting entry, and

- approving service agreements and tariffs that protected pipelines from entry and obliged pipeline customers to pay for the gas reserves acquired on their behalf.[4]

The entire structure of the industry was designed to replace competition with long-term contracts, economic regulation, and restraints on entry, while shifting most of the risk onto end users.[5] Most specifically with regard to acquisition of gas from others, there was no profit opportunity for gas pipelines, but neither was there supposed to be any risk.[6] Stalon and Lock (1990) find

---

[4]   As Kalt (1987, 90), puts it:

> The natural gas industry is subject to considerable risk from both the supply and demand sides. . . . Historically, much of this risk has been channeled to end users in the form of supply shortages. Producers, pipelines, and local distribution companies (distributors) effectively were assured, through regulation and take-or-pay contracts, that they could sell all available gas at determinable prices.

[5]   See Pierce (1982). Like most systems of rigid regulation, "The Act fairly bristles with concern for undue discrimination," *Associated Gas Distributors* v. *FERC*, 824 F.2d (D.C., 1981), suggesting that avoiding discrimination was a higher priority than encouraging competitive pricing.

[6]   As Schuller (1987, 194), puts it:

> In years past, regulation allowed pipelines and distributors to earn acceptable rates of return and serve customers at prices below market levels. These low prices, which create excess demand,

four important consequences of this risk-bearing scheme, with which we believe most observers would concur:

1. The regulatory scheme suppressed competition in favor of creation of monopoly and regulated prices.

2. Producers depended on pipelines to market their gas and distributors depended on pipelines to acquire gas on their behalf.

3. Many of the risks of the interstate pipeline industry were shifted downstream to customers.

4. The regulatory system was almost universally acclaimed (including by financiers who liked the earnings assurances).[7]

THE REALLOCATION OF RISK AFTER THE ELIMINATION OF MINIMUM BILLS

This consensus in support of the old regulatory scheme was lost when low wellhead prices artificially restricted gas output and the willingness of suppliers to sell to the interstate market.   Indeed, the entire system of risk allocation was undermined by a severe

---

allowed regulators to pursue social equity and political goals. This system provided a rate of return for pipelines that covered their capital costs, while certain customers benefitted from stable, below-market prices.

[7]   For discussion, see Stalon and Lock (1990, 480-481). Obviously, producers who were forced to sell gas at levels below market price were not so enamored of the system.

shortage of the vital raw material, cheap gas.[8] To remedy a
shortage of natural gas caused by price controls in the interstate
market, Congress passed the Natural Gas Policy Act (NGPA) in
1978.[9] The series of regulatory and legal events that flowed
from the NGPA is shown in Figure B-2.

Figure B-2
Regulatory Timeline

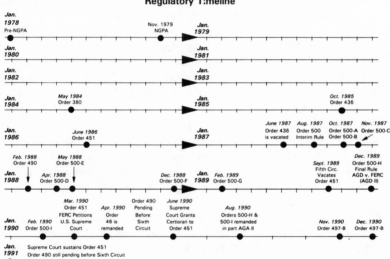

---

[8] Breyer and MacAvoy (1973). Kalt and Schuller (1987, 2) note:

> Before this, the natural gas industry had become accustomed to
> perpetual excess demands. Available supplies virtually were guar-
> anteed to find more than willing buyers, the concepts of aggres-
> sive marketing and strategic planning held little relevance, and
> pervasive regulation had ossified management and contracting prac-
> tices into unchallenged routines.

[9] The Natural Gas Policy Act, Pub. L. No. 95-621, 92 Stat. 3350 (1978)
(codified as amended at 15 U.S.C.A. §§ 3301-3432 (West 1990)).

The NGPA replaced the previous methodology for setting wellhead prices with a complex system of ceiling prices, most of which were to be phased out by 1985.   To encourage production, 'new' gas enjoyed high ceilings.   'Old' gas remained federally regulated. As a result, there were many prices for wellhead gas, depending on vintage and contract terms.   Furthermore, in NGPA Section 311, the Commission was empowered to authorize widespread pipeline transportation without the need for a burdensome proof of public convenience and necessity.

As a result of the NGPA, new gas became available, albeit at a price well above controlled prices of old gas.   Pipelines began to sign long-term contracts for such gas.   Gas supply shortages were perceived to be sufficiently severe that pipelines agreed to stringent take-or-pay provisions in these contracts.   Similar to minimum bills, the take-or-pay provision required the pipeline to pay for a certain fraction of potential deliverability under the contract whether the pipeline needed the gas or not.   Take-or-pay provisions were often for 85 to 90 percent of deliverability, in many cases with prices at the maximum lawful level.

Several factors led pipelines and others to believe that such take-or-pay provisions were sensible.   One was the expectation of rising oil prices, which implied future gas prices would also rise. Another was that the price charged customers was below the contract price because of 'rolled-in pricing.'   Since 'old' gas remained under price controls, the prices customers paid became an average of expensive new gas and cheap old gas, the so-called 'Weighted Average Cost of Gas' (WACOG).[10]   A third, already mentioned, was the hedge provided either by an exclusive service franchise or by minimum bill provisions with pipeline customers.

---

[10]   Note that underpriced gas had been a consistent feature of the prior regulatory scheme.   As Schuller (1987, 185) put it:

> When gas prices were regulated below market prices, marketing consisted simply of order taking and informing interruptible customers when they could expect gas to be turned on or off. [footnote omitted]   In fact, some pipelines abolished their marketing departments altogether.

Unfortunately for the pipeline industry, and much to the surprise of the entire natural gas industry, these factors proved to be insufficient protection.   Oil prices stabilized and eventually collapsed, while average rolled-in gas prices escalated rapidly as a result of terms in the producer contracts.   This led to challenges to pipeline recovery of high-cost gas under the 'fraud and abuse' exception in the NGPA.[11]   Still, from 1983 to 1985, FERC established and permitted a variety of programs designed to maintain pipeline throughput and avoid take-or-pay liabilities.   Many of these 'Special Marketing Programs' (SMPs) involved the abandonment of pipeline sales service, and the direct sale of gas from producers to end users who otherwise might have switched to oil. The pipeline, acting purely as a transportation carrier rather than as a merchant, received take-or-pay relief for the volumes of such sales.

The Commission allowed Special Marketing Programs to lapse after the U.S. Court of Appeals for the D.C. Circuit found them to be unduly discriminatory in *Maryland Peoples Counsel* v. *FERC*, 761 F.2d 768 (D.C. Cir. 1985).   The problem was that the highest cost gas was increasingly being supplied to captive customers who could not switch to oil.   This led to regulatory pressure to make lower cost gas available to captive customers.[12] Moreover, as competition intensified, both the Commission and the gas industry at large began to feel that competition might be a more efficient way to control gas prices than regulatory hearings.

In May 1984, the Commission issued Order No. 380, *Elimination of Variable Costs from Certain Natural Gas Pipeline Minimum*

---

[11]   "[T]he Commission's power to directly restrict pipelines from passing costs through is limited: it can limit pass through of gas purchase costs made in compliance with the NGPA only if the payments arose out of 'fraud or abuse,' NGPA §601(i)(2), 15 U.S.C. §3431(c)(2)(1982), and it can limit pass through of other costs . . . only to the extent not 'prudently' incurred, see C.F.R. §2.76(d)."   *AGD-I* 824 F.2d at 1025.

[12]   See Pierce (1983a) for a litany of complaints about the regulatory and contractual system during this period.

Commodity Bill Provisions,[13] a landmark in gas pipeline history. Order No. 380 eliminated variable, i.e., gas, costs from pipelines' minimum commodity bills, thereby removing the hedge pipelines believed they had for the risk of the long-term supply contracts they had signed.[14] Customers were now free to seek alternative gas supplies at current market prices, which were already below many pipelines' Weighted Average Cost of Gas, and they were not required to pay pipelines for making gas available to them.

The FERC clearly recognized that the effect of Order No. 380 was to shift risk of price variances in gas costs away from downstream customers (distributors and the end users behind them) to the pipelines. The FERC's logic was that this shifting of risk was necessary to achieve the objectives of a more competitive market for wellhead gas and to force gas pipelines to renegotiate their take-or-pay contracts with gas producers:

> The Commission . . . finds that utilization of minimum commodity bills to recover costs for gas not taken is fundamentally inconsistent with the increasingly competitive wellhead market mandated by the Congress in 1978. Congress intended that there be an opportunity for gas prices to increase or decrease -- whichever the market demands. Implementation of the instant rule will further this Congressional intent by removing one obstacle that inhibits response to market demand. [FERC Stats.

---

[13]   Order No. 380, *Elimination of Variable Costs From Certain Natural Gas Pipeline Minimum Commodity Bill Provisions*, 49 Fed. Reg. 22,778 (June 1, 1984), FERC Stats. & Regs. ¶30,571, *reh'g denied and stay granted in part*, Order No. 380-A, 49 Fed. Reg. 31,259 (Aug. 6, 1984), FERC Stats. & Regs. 1982-1985 ¶30,607, *reh'g denied*, Order No. 380-D, 29 FERC ¶61,332 (1984), *aff'd in part, remanded in part sub nom. Wisconsin Gas Co. v. FERC*, 770 F.2d 1144 (D.C. Cir. 1985), *cert. denied sub nom. Transwestern Pipeline Co. v. FERC*, 476 U.S. 1114 (1986), *order on remand.* Order No. 380-E, 35 FERC ¶61,190 (1987).

[14]   For a defense of the efficiency properties of minimum bills, see Masten (1988). For the opposite view, see Legato (1987). See also Hubbard and Weiner (1986).

and Regs., Regulations Preambles 1982-1985, at
30,964]

* * *

The rule will have the effect of allowing certain
natural gas distribution companies to pick and
choose among their pipeline suppliers without incurring charges for gas they do not take. [*Id.* at
30,958]

* * *

*Removal of the insulating minimum commodity bill
. . . places much of the risk of market loss on
the pipeline.* This risk creates an incentive for the
pipeline to go back to its producers and convey
the message that its customers will not buy gas, or
will buy less gas, at such a high price. [*Id.* at
30,963 (emphasis added)]

In short, the whole idea of Order No. 380 is to place more risk
on gas pipelines so that they will be forced to compete and so
that price movements at the wellhead will be felt at the burnertip:

[T]he Commission expects pipelines to make vigor-
ous efforts to compete in the market, which efforts
should result in proper transmittal of price signals
and lower prices for natural gas. (at 30,968)

In AGD-I, 824 F.2d 981 (D.C., 1987), the Court noted:

Besides protecting consumers from the burden of
the pipelines' purchase contracts at supra-market
prices, such rules [allowing customers to buy their
own gas at the wellhead] would have the long-term
effect of subjecting pipelines to the ordinary constraints of a middleman under competitive conditions.

The Commission also required that a portion of the gas released and transported under the SMP be made available to captive customers. Additionally, the Commission began to approve pipeline applications to take on new customers who previously had been served by only one pipeline, thereby eliminating that hedge for the long-term gas supply contracts.[15]

While the FERC was relieving the distributors of their responsibilities for minimum bills, the courts were striking down the transportation programs that denied selective access to depressed spot market gas to many customers.[16] The court was concerned that these transportation programs, approved by the Commission to foster competition, discriminated against customers without alternative fuel choices. Still wishing to establish competition, the Commission in October 1985 issued Order No. 436, *Regulation of Natural Gas Pipelines After Partial Wellhead Decontrol*,[17] which

---

[15] Recent regulatory reforms have created a significant controversy surrounding competition between distributors and pipelines for large industrial customers and new entry into the pipeline industry. See Broadman and Kalt (1989); and MacAvoy, Spulber and Stangle (1989).

[16] *Maryland People's Counsel* v. *FERC*, 761 F.2d 786 (D.C. Cir. 1985) (MPC I); *Maryland People's Counsel* v. *FERC*, 761 F.2d 780 (D.C. Cir. 1985) (MPC II).

[17] Order No. 436, *Regulation of Natural Gas Pipelines After Partial Wellhead Decontrol*, 50 Fed. Reg. 42,408 (October 18, 1985), FERC Stats. & Regs. ¶30,665 (Oct. 9, 1985), *modified*, Order No. 436-A, 50 Fed. Reg. 52,217 (Dec. 23, 1985), FERC Stats. & Regs. ¶30,675 (Dec. 12, 1985), *modified further*, Order No. 436-B, 51 Fed. Reg. 6398 (Feb. 14, 1986), *reh'g denied*, Order No. 436-C, 34 FERC ¶61,404 (Mar. 28, 1986), *reh'g denied*, Order No. 436-D, 34 FERC ¶61,405 (Mar. 28, 1986), *reconsideration denied*, Order No. 436-E, 34 FERC ¶61,403 (Mar. 28, 1986), *vacated and remanded*, *Associated Gas Distributors* v. *FERC*, 824 F.2d 981 (D.C. Cir. 1987). Order No. 500, *Regulation of Natural Gas Pipelines After Partial Wellhead Decontrol*, 52 Fed. Reg. ¶30,334 (Aug. 14, 1987), FERC Stats. & Regs. ¶30,761, *extension granted*, Order No. 500-A, FERC Stats. & Regs. ¶30,770, *modified* Order No. 500-B, FERC Stats. & Regs. ¶30,772, *modified further*, Order No. 500-C, FERC Stats. & Regs. ¶30,786 (1987), *modified further*, Order No. 500-D, FERC Stats. & Regs. ¶30,800, *reh'g denied*, Order No. 500-E, 43 FERC ¶61,234, *modified further*, Order

required gas pipelines to provide transportation services on a non-discriminatory basis, eliminating the ability of any pipeline to favor transportation of its own gas over transportation of independently purchased gas. Order No. 436 and its progeny were extremely complex in their implementation, but their consequences for pipeline risk are clear.

Relieved of their contract obligations to buy pipeline tariff gas by Order No. 380, customers were now as a result of Order No. 436 in a position to 'unbundle' the gas transportation function from the gas acquisition function. Once a pipeline accepted an open-access blanket certificate, the effect of Order No. 436 was to put the pipeline's merchant function in direct competition with its transportation function, as well as with other pipelines and other fuels. Meanwhile, Order No. 436 did not explicitly provide for a mechanism for gas pipelines to deal with their take-or-pay gas contracts. Indeed, this was one of the chief factors considered by the D.C. Circuit when it remanded Order No. 436 back to the FERC for reconsideration.[18]

No. 500-F, FERC Stats. & Regs. ¶30,841 (1988), *reh'g denied*, Order No. 500-G, 46 FERC ¶61,148 (1989), *vacated and remanded, American Gas Association v. FERC*, 888 F.2d 136 (D.C. Cir. 1989) (AGA-I). Order No. 500-H, Final Rule, FERC Stats. & Regs. ¶30,867 (Dec. 13, 1989). Order No. 500-I, FERC Stats. & Regs. ¶30,880 (Feb. 12, 1990), *aff'd in part and remanded in part, American Gas Association v. FERC*, 912 F.2d 1496 (D.C. Cir. 1990) (AGA-II).

[18]   *Associated Gas Distributors* v. *FERC* (AGD-I), 824 F.2d 1023-25 (D.C., 1987). One of the chief concerns of the court was that Order No. 436 relieved downstream customers of obligations to purchase pipeline gas but did not adequately address the effect of open access on pipeline take-or-pay obligations:

> . . . the Order effectively reduces pipeline ability to face down recalcitrant producers. . . . Customers' conversion to transportation will clearly aggravate a pipeline's ability to resolve the problem of its overpriced gas inventory.
>
>   *   *   *
>
> In sum, FERC's seeming blindness to the possible impact of Order 436 on take-or-pay liability, and its tendency to elevate into

Order No. 436 also attempted to increase 'gas on gas' competition by encouraging new entry.   A provision for 'optional expedited certificates' sought to stimulate new entry by cutting through the cumbersome process of obtaining FERC approval for new facilities or for expansion of service.   FERC approval has involved the issuance of certificates of public convenience and necessity, typically a long and costly process including studies of projected impacts on shareholders and customers of both the applicant firm and of any firm threatened by displacement of sales.[19]

Order No. 436 sought to expedite the application process by granting applicants rebuttable presumptions of public convenience and necessity if they met certain criteria.   The criteria were designed to place on applicants all the risk that the expanded facilities or service might not eventually be either 'convenient' or 'necessary.'[20]   For example, one of the criteria was that the applicant waive the ability to adjust projected volumes downward in future rate cases.   Applicants for 'optional expedited certificates' would therefore be ineligible for rate relief from unanticipated slacks in business volumes.[21]

In an attempt to bring supply contract prices more in line with the prevailing market price of gas, the Commission issued Order No. 451, *Ceiling Prices: Old Gas Pricing Structures,*[22] consolidating the various classifications of old gas into a single group with one ceiling price, which happened to be above market-clearing levels.   Once producers invoke negotiation under this order,

---

affirmative benefits what are at best palliatives, seem impossible to square with the requirement of reasoned decision-making.

[19]   *Associated Gas Distributors* v. *FERC*, 824 F.2d at 1030 (D.C. Cir. 1987).

[20]   *Id.* at 1038.

[21]   *Id.* at footnote 32.

[22]   Order No. 451, *Ceiling Prices: Old Gas Pricing Structure,* III F.E.R.C. Stats. & Regs. ¶30,701, 51 Fed. Reg. 22,168 (1986).   See Shoneman and McConnell (1986).

pipelines must either abandon the contract or pay much higher rates. In *Mobil Oil Exploration and Producing Southeast, Inc. et al.* v. *FERC* 885 F.2d 209 (5th Cir. 1989), the 5th Circuit vacated Order No. 451 on the grounds that, among other things, the FERC had exceeded its authority to raise prices, exacerbated the take-or-pay problem and improperly failed to confront the take-or-pay problem. The Supreme Court, however, reversed the 5th Circuit on January 8, 1991,[23] so Order No. 451 is finally firm.

Additionally, Order No. 490 permits producers and open-access pipelines to abandon low-priced sales to pipelines when the contracts expire.[24] However, after five years FERC has still not resolved the issue of 'pre-granted abandonment' that would permit pipelines to abandon service with their customers upon the expiration of the agreement.[25]

Finally, the Natural Gas Wellhead Decontrol Act of 1989 deregulates all producer prices as of January 1, 1993.[26] Pipelines therefore will have to renegotiate producer contracts, perhaps at sharply higher prices.

---

[23]   59 LW 4054-60.

[24]   *Abandonment of Sales and Purchases of Natural Gas Under Expired, Terminated, or Modified Contracts*, FERC Stats. and Regs. ¶30,797, reh'g denied, Order No. 490-A, FERC Stats. and Regs. ¶30,825 (1988).

[25]   Thus as Pierce (1990, 10) puts it:

> To illustrate the present situation with an extreme example, assume that a pipeline has aggregate certificated sales obligations to its distributor customers of one billion cubic feet (Bcf) per day. Assume further that distributors choose to buy no gas from the pipeline for a period of ten years. At the end of that ten-year period, each distributor notifies the pipeline that it desires to buy its full certificated volume from the pipeline. Unless FERC previously had authorized abandonment of the pipeline's sales service, the pipeline would have a regulatory obligation to supply one Bcf of gas per day immediately.

[26]   Pub. L. No. 101-60, 103 Stat. 157 (1989).

EFFECTS OF OPEN ACCESS TRANSPORTATION ON RISK ALLOCATION

Order No. 436 not only allowed pipeline customers to displace pipeline merchant gas (i.e., gas owned and transported by the pipeline) with transportation gas; it also enabled customers to reduce their contract demand levels with pipelines to lower demand charge payments. This provision of Order No. 436 was later overturned by the U.S. Court of Appeals, D.C. Circuit (AGD-I, 824 F.2d 981), but the court left standing a provision that enabled pipeline customers to convert up to 100 percent of their sales volumes to firm transportation.

The Court in AGD-I was particularly concerned that the FERC had made no explicit effort to address the take-or-pay problem. Order No. 380 relieved pipeline customers of minimum bills, but pipelines were not given similar authority to modify their service obligations to customers. Nor, more importantly, were they explicitly given a way to modify their contracts with producers. Thus pipelines were left with long-term contracts for gas, some at very high prices, and no way to pass all of these prices through to customers. The pipelines' hedges against risks of gas acquisition had simply failed.

In an attempt to remedy the Court's concerns with Order No. 436, the FERC published Order No. 500 (August 7, 1987) and its progeny (the '500 series'). Order No. 500 addressed many of the issues that the Court found most troublesome in Order No. 436. First, it proposed a 'sharing' mechanism for allocating the take-or-pay liability among the various parties. A pipeline may recover take-or-pay costs through its commodity charges (but, of course, market forces may prevent this if the pipeline is an open-access transporter). An open-access transporter could charge 25 to 50 percent of its take-or-pay costs in a fixed charge if it agrees to absorb an equal amount. It may then attempt to recover the remainder through sales and transportation surcharges.[27]

Second, to prevent recurrence of take-or-pay obligations, the Commission outlined a policy that would permit pipelines to levy a Gas Inventory Charge. Any pipeline with a Gas Inventory Charge waives the right to any other take-or-pay cost recovery mechanism for future obligations.

Third, to mitigate the effects of open access transportation on take-or-pay liabilities, the FERC created a new crediting mechanism. Under the rule, a gas producer seeking open access transportation would have to permit pipelines to credit transported gas against any take-or-pay liability that arose for that pipeline from that producer in another contract.[28]

The '500' series of orders retained the mechanism whereby gas pipeline customers could convert rights to sales volumes to rights to transportation capacity, but dropped the provision whereby customers could reduce the sales rights with no conversion.

In response to Order No. 500, the Court of Appeals again remanded the decision on the grounds that Order No. 500 did not

---

[27]   In response to a court of appeals decision, the Commission has stayed the authority of some pipelines to collect fixed charges which were based on the purchase deficiency allocation method. The Commission also has recognized that by using a different allocation mechanism, pipelines "may seek to recover take-or-pay costs that were previously being recovered under a fixed charge mechanism that has been invalidated by the court of appeals." Order No. 528, "Order on Remand Staying Collection of Take-or-Pay Fixed Charges and Directing Filing of Revised Tariff Provisions," issued November 1, 1990, in Docket No. RM91-2-000, *et al.*, *Mechanisms for Pass through of Pipeline Take-or-Pay Buy out and Buy down Costs*, 53 FERC ¶61,163, at 61,595 (1990).

[28]   INGAA (1989) reports that "this crediting feature had little effect on the reduction of take-or-pay exposure levels. . . ."

comply with the AGD-I decision.[29]   The Court stated that the
FERC ". . . appears at some point to have abandoned the effort
to produce a final rule, so satisfied is it with the status quo."

Pipelines, therefore, have been forced to buy out of many of
their contracts with producers, or to litigate these contracts, at a
cost of billions of dollars.   Table B-1 shows the take-or-pay
exposure of the interstate pipelines.   FERC estimates that the
pipelines have had to absorb $3.4 billion of the take-or-pay prob-
lem.[30]

The central theme of these developments is one of great uncertainty.   The effect of each of the major regulatory developments

---

[29]   In *American Gas Association et al.* v. *FERC* (AGA-I), 888 F.2d 136 (D.C.
Cir. 1989), the Court remanded Order No. 500-A through G to the FERC.
In 'AGA-II,' the Court ruled that Order No. 500-H and Order No. 500-I
complied with AGD-I, but remanded the issues of pre-granted abandonment
and double crediting. *American Gas Association* v. *FERC*, 912 F.2d 1496
(D.C. Cir. 1990) (AGA-II).

[30]   Because of the specific mechanisms in individual contracts, it is extremely
difficult to obtain accurate estimates of the actual liabilities arising from
take-or-pay contracts.   FERC arrived at the $3.4 billion figure as follows
in Order No. 500-H, pp. 44-45:

> Under Order No. 500's equitable sharing mechanism, the pipelines
> are absorbing 39.3 percent of the $8.6 billion in payments to
> producers included in Order No. 500 filings or about $3.4 billion,
> while recovering through a fixed take-or-pay charge another 39.3
> percent, and assessing the remainder through volumetric surcharges
> to both sales and transportation customers.   [Footnote omitted.]
> In addition, pipelines may recover the part to be recovered
> through the volumetric surcharge only to the extent market forces
> permit them to raise their prices sufficiently to recover the volu-
> metric surcharge.

Pipelines have make-up periods to take the gas already paid for and many
contracts have been renegotiated.   INGAA estimates that, as of September
30, 1989, there were 19 pipelines with $2.3 billion in remaining TOP
'exposure.'   Actual payments for 'buy out/buy down' represent some fraction of that amount, which INGAA (1989) reports has increased from 11
percent in 1985 to 37 percent in the second and third quarters of 1989.

			TABLE B-1 TAKE-OR-PAY EXPOSURE OF THE INTERSTATE PIPELINES $ BILLIONS				
	1984	1985	1986	1987	1988	1989
Obligation Start of Year		$4.7	$6.1	$10.0	$8.2	$3.8
Accumulated During Year		$5.5	$6.1	$5.5	$3.1	$1.0
Settled with Producers		$3.7	$5.4	$8.1	$7.5	$2.8
Statistical Differences*		($0.5)	$3.2	$0.8	$0.0	$0.3
Obligation End of Year	$4.7	$6.1	$10.0	$8.2	$3.8	
Obligation September 30						$2.3
TOTAL ACCUMULATED (includes Adj.)		$29.8				
TOTAL SETTLEMENTS		$27.5				
Cost of Settlement			$8.8			

*Change in Sample Size, Retroactive Adjustments
Source: Pipeline surveys by Interstate Natural Gas Association of America, as cited in Jensen Associates
(1990).

has been to shift risk back upstream and away from consumers,
after the magnitude of the downside risks became clear.   Considerable time passed as these events unfolded, and almost every
important FERC decision has been remanded by an appeals court,
creating further uncertainty and delay.[31]

---

[31]   In *Illinois, ex rel. Hartigan* v. *Panhandle Eastern Pipe Line Co.*, 1990-1
Trade Cases, 62,683; 62,726; 62,727; the court found that ". . . in the
period of time of 1981 to 1986, nobody knew with any certainty where
and under what conditions the natural gas market and FERC regulations
would finally come to rest."   It referred to ". . . the chaos in the marketplace and the uncertainty regarding the ultimate shape of new regulations
by FERC . . . ."   *The Wall Street Journal* (Soloman and Solis 1989, 1)
described the history even more forcefully:

RATE DESIGN AND RISK ALLOCATION

Gas historically was purchased as a bundled combination of transportation and the gas commodity, collectively called 'tariff gas,'
or as gas supply along with an unbundled transportation service
from a third-party pipeline.   In both cases, the rate was usually
a two-part rate composed of a *demand charge* and a *commodity
charge* (see Figure 6-1).   The demand charge is often thought of
as a capacity charge because it is based on the right to take a
maximum volume during the peak demand periods.   The commodity charge is based on the actual volume of gas taken during the
time period.[32]

Because different classes of customers have different load factors
(i.e., actual usage of capacity, which varies generally because of
seasonal variability), these rate structures have become the focus
of considerable dispute among ratepayers.   They also have considerable implications for efficient use of gas and pipeline capacity.

The easiest way to see the significance of rate design for allocation of risks in the pipeline industry is to start with a pure
fixed-variable rate design.[33]   This approach puts the responsibility
for recovering all fixed ('nontraffic sensitive' in economic terms)
costs into the demand charge and all variable costs into the commodity charge.   It might appear initially that this result would be
most economically efficient (i.e., it purports to set incremental
price to incremental cost) but the pure fixed-variable rate design
has been modified considerably in practice.   The reasons usually
cited are that the pipeline is indifferent to throughput (and thus

---

Regulators have changed the rules for the umpteenth time.   Bil-
lion-dollar litigation understood by only a handful of lawyers has
crippled balance sheets.

[32]   For more details, see American Gas Association (1987).

[33]   This approach, while sometimes thought to absolve pipelines of too much
risk in this country, is used in Canada.   TransCanada Pipelines is currently
on a pure fixed-variable rate design.   See McGee and Lamphier (1990).

allegedly insulated from risk) and that such a rate structure does not necessarily price gas at correct levels because the gas itself may be underpriced. Loading some fixed costs into the commodity charge also responds to regulatory concern that interruptible customers make some contribution to fixed costs.

In 1952, the Federal Power Commission employed the 'Seaboard' rate design, whereby fixed costs were allocated 50/50 between the demand and commodity components.[34] In response to emerging shortages of natural gas arising from price controls, the FERC switched to the 'United' scheme, whereby 75 percent of fixed costs was in the commodity charge.[35] Later, when the problem was viewed as overpriced gas, the FERC switched to the modified fixed-variable (MFV)[36] scheme under which the return on equity and related tax components, variable costs, and all production and gathering costs were left in the commodity charge. To take some of the edge off the shift in cost recovery toward the low load factor customers (chiefly households and commercial establishments who use gas for heating) resulting from higher demand charges, the higher demand charge was split into two parts, one of which depended on annual usage and the other on peak usage.[37]

---

[34]   *Atlantic Seaboard Corporation*, 11 FPC 43 (1952).

[35]   *United Gas Pipe Line Corporation*, 50 FPC 1348 (1973).

[36]   MFV was first adopted in 1983 in *Natural Gas Pipeline Company of America*, 25 FERC ¶61,176 (1983), *order on reh'g*, 26 FERC ¶61,203 (1984), *aff'd in relevant part*, *Northern Indiana Public Service Co. v. FERC*, 782 F.2d 730 (7th Cir. 1986).

[37]   Half of the costs assigned to the demand charge are recovered by a rate based on peak usage (D-1) and half is based on annual usage (D-2).

On May 30, 1989, the FERC issued its *Policy Statement Providing Guidance with Respect to the Designing of Rates.*[38]   The objective of these proposals is to make rate structures more responsive to economic efficiency considerations.   No firm requirements were offered, but the Commission contemplated the possibility of changes in the areas of seasonal rates and a shift away from use of the D-2 component.[39]   The Commission considered the issue of discounting, which is the freedom of pipelines to price within a range of minimum and maximum set rates.   Proposals had been made to incorporate projected discounting into the volume assumptions when determining maximum rates.   The Commission rejected this approach, recognizing that pipeline discounting would be deterred if the volumes obtained through this process could subsequently be used against pipelines in the rate-setting process.[40]   Recommendations were made to keep pipelines from being placed at risk for discounted volumes.

Notably, the chief focus of the efficiency concerns was to maximize pipeline throughput and to allocate scarce peak-period capacity efficiently among users.   The efficient allocation of risk between pipelines and customers was not explicitly addressed.   However, the policy would appear to give greater scope for pipelines to use rate structures as a tool for competition.

The essential point to recognize in this history is that these changes in rate structure were not cost-driven.   Rather, they were responsive to changes in the demand environment (chiefly the marketability of the gas, given the commodity charge).   'Everything else' is decidedly not equal as one moves from the environment of one rate design scheme to the other.

---

[38]   *Interstate Natural Gas Pipeline Rate Design, et al.*, 47 FERC ¶61,295, *order on reh'g* 48 FERC ¶61,122 (1989), *appeal docketed, Wisconsin Public Service Commission v. FERC*, No. 89-1598 (D.C. Cir. September 25, 1989).   See also Office of Economic Policy (1989) and Horvath (1989b).

[39]   See footnote 37 for the definition of the D-2 charge.

[40]   *Interstate Natural Gas Pipeline Rate Design, et al.*, pp. 13-16.

GAS INVENTORY CHARGES (GICs)[41]

On May 30, 1989, the FERC published its notice of proposed policy statement on so-called Gas Inventory Charges, to clarify its initial enunciation of the concept in Order No. 500.[42] The objective of such charges is to compensate pipelines for the risk they bear from being required to stand ready to sell the pipeline's tariff gas while giving customers the option to transport third-party gas whenever that is cheaper. Gas Inventory Charges would not apply for past take-or-pay obligations.

The need for a GIC arises because pipelines are at risk for a mismatch between the gas for which they contract and the gas their customers decide to take. For example, if the pipeline contracts for gas based on past takes by the customer and/or its service obligation and the customer decides to switch to firm transportation, the pipeline is at risk. Economically, the GIC is an 'insurance premium' against such risks.

The discussion of Order No. 490 above illustrates that risk might arise on the other side, as well. If the pipeline does not have a ready (or perhaps in some circumstances, *any*) mechanism to eliminate its service obligation even if the customer has not been taking the pipeline's gas for some time, and the customer announces it now plans to buy a large amount, the pipeline may incur losses while scrambling to find a supply or may be penalized in a prudence hearing for not anticipating the customer's switch and contracting for more standby gas.

Since customers are most likely to change previous take levels when the price of spot gas changes dramatically, giving customers control of the process creates risk for the pipeline. The GIC in

---

[41] We are indebted to Frank Graves and George Hall for extensive discussions of the theory and current practice of GICs. Of course, we retain responsibility for any defects in the following discussion.

[42] FERC, *Interim Gas Supply Charges and Interim Gas Inventory Charges*, 47 FERC ¶61,294 (1989).

principle can compensate for this risk, but the difference between principle and practice may well turn out to be important.   The Gas Inventory Charge rate-setting process is complicated and not yet well defined in practice.   Two issues arise for a given pipeline's GIC: (1) what standard the FERC uses to determine whether a proposed GIC is reasonable; and (2) how the GIC payment itself is made by pipeline customers.   We discuss these issues in turn.

The standard the FERC uses is either 'market-based' or 'cost-based.'   If the FERC determines that the market the pipeline serves has sufficient competition, the pipeline can set a market-based GIC at whatever level it chooses.   If some competition exists, but at an insufficient level, the GIC must be cost-based. Also, a third form, an 'as-billed' GIC, has been proposed for downstream pipelines who want to recover the cost of GICs charged by upstream pipelines.   We understand that most pipelines that have requested GICs so far have asked for a cost-based GIC, but hard data on this issue are unavailable.

The problem, then, is how to define the 'cost' of the risk to which the pipeline is exposed.   Various rules of thumb have been proposed, but to our knowledge, no one to date has proposed a GIC based on a serious analytical attempt to assess the distribution of possible losses and to find an actuarially fair insurance premium based on that distribution (although no one can be sure of what approach might have been taken within individual companies).

A second problem with a cost-based GIC is that after some period of time, pipelines are supposed to provide refunds to their customers of 'excess' GIC charges.   The formula by which these refunds will be allocated remains undefined.   Apparently, the refund policy has arisen because of uncertainty about what an actuarially fair cost-based GIC would be, and the notion of refunds is that it protects customers if the early rules of thumb turn out to yield GICs that are too high.   But the insurance premium analogy points to serious problems with a refund policy. Should a life insurance company, for example, be required to refund premiums to those who *did not* die during part of the life of the policy? Then where is the money to come from to pay the death benefit when the policyholder *does* die? If GICs are truly to provide adequate risk compensation to pipelines in the long run, practical problems of this sort will have to be solved.

The second issue that arises for a particular GIC is how the customers will pay for it. Here two options exist, 'demand' and 'deficiency.' The first is a two-part charge, with a commodity charge based on an index of competitive market prices and a demand charge that includes a premium to have gas available at the commodity rate. Under the deficiency method, the customer pays for each unit of gas it fails to take below a specified level. In practice, cost-based and deficiency GICs seem to go together and market-based and demand GICs seem to go together, although this is not strictly a requirement. So far, deficiency GICs have used a single charge for the gas not taken, although there might be some useful incentives if the deficiency penalties increased as the customer's take fell further below the agreed upon minimum.

In sum, while the outlines for a system of GIC payments to compensate pipelines for bearing gas contracting risks are in place, many details of implementation remain to be worked out. Moreover, there is no guarantee that the ultimate implementation procedures will produce GICs that actually correspond to the risks pipelines bear. Indeed, there appears to be little effort aimed at this seemingly vital question.

CONCLUSION

The integrated gas industry from wellhead to burnertip now experiences considerably more risk than during the period when abundant supplies of gas were available at below-market prices. Natural gas has become another commodity, with all the attending risks of price fluctuations. Furthermore, these risks of price variations imply additional risk up and down the pipeline system. In response to this greater risk, regulatory and contractual changes have shifted this risk from customers back upstream to pipelines and producers. Unless that additional risk is wholly shifted by the pipeline back to producers, which appears to be infeasible in the current regulatory and contractual environment, pipelines clearly have additional risk over that experienced in the prior regulatory regime. The issue to be resolved in the body of this book is whether this degree of risk may nonetheless be characterized as 'low' relative to the recent past or to other industries.

Bibliography

Ahn, Chang Mo and Howard E. Thompson. 1989. "An Analysis of Some Aspects of Regulatory Risk and the Required Rate of Return for Public Utilities." *Journal of Regulatory Economics* 1: 241-257.

American Bar Association. 1990. *Newsletter* (Public Utility Section). January.

American Gas Association. 1983. *The Changing Nature of Investment Risk for the Natural Gas Industry*. Arlington, Va.: AGA.

_____. 1987. *Gas Rate Fundamentals*. Fourth Edition. Arlington, Va.: AGA.

Arthur K Worldwide Organization/Cambridge Energy Research Associates. 1991. *Natural Gas Trends 1991: A Statistical Profile*. Chicago, Ill.: Arthur Andersen & Co.; Cambridge, Mass.: CERA.

Averch, H. and L. L. Johnson. 1962. "Behavior of the Firm under Regulatory Constraint." *American Economic Review* 52: 1052-1069.

Bailey, Elizabeth E. and John C. Malone. 1970. "Resource Allocation and the Regulated Firm." *Bell Journal of Economics* 1: 129-142.

Baron, D. P. 1988. "Regulation and Legislative Choice." *RAND Journal of Economics* 19 (Autumn): 467-477.

Baumol, William J. and Alvin K. Klevorick. 1970. "Input Choices and Rate-of-Return Regulation: An Overview of the Discussion." *Bell Journal of Economics* 1: 162-190.

Beranek, William and Keith M. Howe. 1990. "The Regulated Firm and the DCF Model: Some Lessons from Financial Theory." *Journal of Regulatory Economics* 2: 191-200.

Berlin, Edward. 1984. "Excess Capacity, Plant Abandonments and Prudence: The Appropriate Regulatory Standard." *Public Utilities Fortnightly*, November 22, pp. 26ff.

Blackmon, Glenn and Richard Zeckhauser. 1992. "Fragile Commitments and the Regulatory Process." *Yale Journal on Regulation* 9 (Winter): 73-105.

Bonbright, James C. 1961. *Principles of Public Utility Rates*. New York: Columbia University Press.

Brealey, Richard A. and Stewart C. Myers. 1988. *Principles of Corporate Finance*. 3rd ed. New York: McGraw-Hill).

Breyer, S. and Paul MacAvoy. 1973. "The Natural Gas Shortage and the Regulation of Natural Gas Producers." *Harvard Law Review* 86: 941-987.

Broadman, Henry G. and Joseph P. Kalt. 1989. "How Natural is Monopoly? The Case of Bypass in Natural Gas Distribution Markets." *Yale Journal on Regulation* 6 (Summer): 181-208.

Bryan, William Jennings. 1906. "Introduction." In *The World's Famous Orations*, edited by William Jennings Bryan. New York: Funk and Wagnalls Company.

Buchmann, Alan P. 1991. "The *Duquesne* Opinion: A Practitioner's Comment." *Yale Journal on Regulation 8 (Winter): 165-170.*

Carpenter, Paul R., Henry D. Jacoby, and Arthur W. Wright. 1987. "Adapting to Change in Natural Gas Markets." In *Energy: Markets and Regulation*, edited by Richard L. Gordon, Henry D. Jacoby, and Martin B. Zimmerman. Cambridge, Mass.: The MIT Press.

Cook, James. 1985. "Nuclear Follies." *Forbes*, February 11, p. 82.

Copeland, Basil L., Jr. and Walter W. Nixon, III. 1991. "Procedural Versus Substantive Economic Due Process for Public Utilities." *Energy Law Journal* 12: 81-110.

Council on Economic Regulation. 1989. *Regulation and Competition in Natural Gas Markets*. Washington, D.C.: CER, September.

Cudahy, Judge Richard D. 1991. "Comment: Shedding Light on *Duquesne.*" *Energy Law Journal* 12: 259-264.

Darr, Frank. 1990. "The Constitutional Limits on Ratemaking: A Response to William [*sic*] Pond." *Energy Law Journal* 11: 53-64.

Drobak, John N. 1985. "From Turnpike to Nuclear Power: The Constitutional Limits on Utility Rate Regulation." *Boston University Law Review* 65: 65-125.

Electric Utility Week. 1985. "PS Indiana to Minimize New Plant Spending until State Changes Attitude." April 29, p. 3.

Etzioni, Amitai. 1986. "Does Regulation Reduce Electricity Rates? A Research Note." *Policy Sciences* 19: 349-357.

Federal Energy Regulatory Commission. 1989. *Policy Statement Providing Guidance with Respect to the Designing of Rates*. May.

Ferris, S. P., D. J. Johnson, and D. K. Shome. 1986. "Regulatory Environment and Market Response to Public Utility Rate Decisions." *The Journal of Finance Research* (Winter): 313-318.

Goldberg, Victor P. 1976. "Regulation and Administered Contracts." *Bell Journal of Economics* 7 (Autumn): 426-448.

Goldsmith, Richard. 1989. "Utility Rates and 'Takings.'" *Energy Law Journal* 10: 241-276.

Graubard, Stephen R. 1990. "Preface" to Issue on Risk. *Daedalus* 119 (Fall): v-vi.

Greenhouse, Linda. 1989. "High Court Rejects Charges by Utilities for Unused Plants." *New York Times*, January 12, p. A1.

Hall, George R. 1987. "Getting Regulation from 'Here' to 'There.'" In *Drawing the Line on Natural Gas Regulation*, edited by Joseph P. Kalt and Frank C. Schuller. New York: Quorum Books.

Hays, William L. and Robert L. Winkler. 1970. *Statistics*. Vol. I. New York: Holt, Rinehart, and Winston.

Hoecker, James J. 1987. "'Used and Useful': Autopsy of a Ratemaking Policy." *Energy Law Journal* 8: 303.

Hogan, William W. and A. Lawrence Kolbe. 1991. "How Far Back Should Prudence Tests Reach?" *Public Utilities Fortnightly* 127 (January 15): 34-37.

Horvath, R. Skip. 1989a. "Marketplace Needs vs. Current Systems." *Natural Gas* (April): 21-23.

_____. 1989b. "Economics of the Commission's Rate-Design Policy Statement." *Natural Gas* 6 (September): 1-7.

Hubbard, R. Glenn and Robert J. Weiner. 1986. "Regulation and Long-Term Contracting in U.S. Natural Gas Markets." *The Journal of Industrial Economics* 35 (September): 71-79.

Ibbotson Associates, Inc. 1991. *Stocks, Bonds, Bills and Inflation, 1991 Yearbook.* Chicago, Ill.: R. G. Ibbotson Associates, March.

Interstate Natural Gas Association of America (INGAA). 1988. *Discounted Transportation on Interstate Natural Gas Pipelines. No. 88-10.* Washington, D.C.: INGAA, November.

_____. 1989. *Take-or-Pay Exposure and Costs Through September 30, 1989.* No. 89-8. Washington, D.C.: INGAA.

_____. 1990. *December 1989 Events Survey.* Washington, D.C.: INGAA.

_____. 1991a. *Carriage Through 1990.* No. 91-2. Washington, D.C.: INGAA, April.

_____. 1991b. *The Costs of Implementing Comparability.* No. 91-3. Washington, D.C.: INGAA, July.

Jensen Associates. 1990. *U.S. Open Access Gas Pipeline Transportation -- A Model for Europe?* Boston: Jensen Associates, September.

Johnson, Robert. 1992. "Fueling Anger: Natural Gas Prices Soak Both Producers, Users: Are Utilities at Fault?" *The Wall Street Journal*, April 2, p. A1.

Joskow, Paul L. 1974. "Inflation and Environmental Concern: Structural Change in the Process of Public Utility Price Regulation." *Journal of Law and Economics* 17: 291-327.

Kahn, Alfred E. 1970. *Economics of Regulation*. Vol. I. New York: John Wiley and Sons, Inc.

_____. 1985. "Who Should Pay for Power Plant Duds?" *The Wall Street Journal*, August 15.

Kalt, Joseph P. 1987. "Market Power and the Possibilities for Competition." In *Drawing the Line on Natural Gas Regulation*, edited by Joseph P. Kalt and Frank C. Schuller. New York: Quorum Books.

Kalt, Joseph P. and Frank C. Schuller, editors. 1987a. *Drawing the Line on Natural Gas Regulation*. New York: Quorum Books.

_____. 1987b. "Introduction: Natural Gas Policy in Turmoil." In *Drawing the Line on Natural Gas Regulation*, edited by Joseph P. Kalt and Frank C. Schuller. New York: Quorum Books.

Kalt, Joseph P., Henry Lee, and Herman B. Leonard. 1987. "Re-Establishing the Regulatory Bargain in the Electric Utility Industry." Energy and Environmental Policy Center, John F. Kennedy School of Government, Harvard University, Discussion Paper No. E-87-02.

Kaplow, Louis. 1989. "An Economic Analysis of Legal Transitions." *Harvard Law Review* 99: 509.

Kinsey, Tim. 1990. "The Credit Quality of the Natural Gas Industry." Working Paper, presented at Proceedings of the Seventh

NARUC Biennial, Columbus, Ohio, NRRI Conference, September.

Knight, Frank H. 1921. *Risk, Uncertainty, and Profit.* Boston: Houghton Mifflin Company.

Knight, Franklin D. 1991. *Testimony Provided in the Matter of Texas Gas Transmission Corporation.* Docket No. RP 88-115-000 *et al.* July 10.

Kolbe, A. L. and R. W. Hodges. 1989. *EPRI PRISM Interim Report: Task 5 -- Other Industry Survey, Parcel/Message Delivery Services.* EPRI Research Project RP-2801-2. Palo Alto: EPRI, June 20.

Kolbe, A. Lawrence and James A. Read, Jr., with George R. Hall. 1984. *The Cost of Capital.* Cambridge, Mass.: MIT Press.

Kolbe, A. Lawrence and William B. Tye. 1991. "The *Duquesne* Opinion: How Much *'Hope'* Is There for Investors in Regulated Firms?" *Yale Journal on Regulation* 8 (Winter): 113-157.

_____. 1992. "Environmental Cleanup Liabilities." *Public Utilities Fortnightly* 129 (January 1): 18-20.

Kolbe, A. Lawrence, William B. Tye, and Miriam A. Baker. 1984. "Conditions for Investor and Customer Indifference Among Regulatory Treatment of Deferred Income Taxes." *Rand Journal of Economics* 15 (Autumn): 434-446.

Kolbe, A. Lawrence, William B. Tye, and Stewart C. Myers. 1991. *Risk of the Interstate Natural Gas Pipeline Industry.* Washington, D.C.: Interstate Natural Gas Association of America.

Lee, Wayne L. and Anjan V. Thakor. 1987. "Regulatory Pricing and Capital Investment Under Asymmetric Information about Cost." *Southwestern Economic Journal* 53 (January): 720-734.

Legato, Carmen. 1987. "The Role of Regulation in Risk Allocation." In *Drawing the Line on Natural Gas Regulation*, edited by Joseph P. Kalt and Frank C. Schuller, 207-233. New York: Quorum Books.

Lyon, Thomas P. 1991. "Regulation with 20-20 Hindsight: 'Heads I Win, Tails You Lose'?" *RAND Journal of Economics* 22 (Winter): 581-595.

MacAvoy, Paul W., Daniel F. Spulber, and Bruce E. Stangle. 1989. "Is Competitive Entry Free? Bypass and Partial Deregulation in Natural Gas Markets." *Yale Journal on Regulation* 6 (Summer): 209-248.

Masten, Scott E. 1988. "Minimum Bill Contracts: Theory and Policy." *Journal of Industrial Economics* 37 (September): 85-97.

McGee, Suzanne and Gary Lamphier. 1990. "Recession-Resistant TransCanada Pipelines is Poised to Realize Strong Growth Potential." *The Wall Street Journal*, November 6.

Meyer, John R. and William B. Tye. 1988. "Toward Achieving Workable Competition in Industries Undergoing a Transition to Deregulation: A Contractual Equilibrium Approach." *Yale Journal on Regulation* 5 (Summer): 273-297.

Myers, Stewart C. 1972. "Application of Finance Theory to Public Utility Rate Cases." *Bell Journal of Economics* 3 (Spring): 58-97.

_____. 1988. "Fuzzy Efficiency." *Institutional Investor* (December): 8-9.

Myers, Stewart C., A. Lawrence Kolbe, and William B. Tye. 1984. "Regulation and Capital Formation in the Oil Pipeline Industry." *Transportation Journal* 23 (Summer): 25-49.

_____. 1985. "Inflation and Rate of Return Regulation." *Research in Transportation Economics* 2: 83-119.

Office of Economic Policy. 1989. *Efficient Rate Designs for Natural Gas Pipelines*. Washington, D.C.: OEP, September.

Peltzman, Sam. 1976. "Toward a More General Theory of Regulation." *Journal of Law and Economics* 19 (August): 211-240.

Pierce, Richard J., Jr. 1982. "Natural Gas Regulation, Deregulation, and Contracts." *Virginia Law Review* 68 (January): 63-115.

_____. 1983a. "Reconsidering the Roles of Regulation and Competition in the Natural Gas Industry." *Harvard Law Review* 97 (December): 345-385.

_____. 1983b. "The Regulatory Treatment of Mistakes in Retrospect: Canceled Plants and Excess Capacity." *The University of Pennsylvania Law Review* 132 (December): 497-560.

_____. 1988. "Reconstituting the Natural Gas Industry from Wellhead to Burnertip." *Energy Law Journal* 9: 1-58.

_____. 1989. "Public Utility Regulatory Takings: Should the Judiciary Attempt to Police the Political Institutions?" *The Georgetown Law Journal* 77 (August): 2031, 2050-2051.

_____. 1990. *Regulation and Competition in Natural Gas Distribution*. Washington, D.C.: Council on Economic Regulation, July.

_____. 1991. "The Unintended Effects of Judicial Review of Agency Rules: How Federal Courts Have Contributed to the Electricity Crisis of the 1990s." *Administrative Law Review* 43 (Winter): 7-29.

_____. 1992. "Placing the *Duquesne* Opinion in a Political Framework." *Research in Law and Economics* 15. Forthcoming.

Pond, Walter. 1989. "The Law Governing the Fixing of Public Utility Rates: A Response to Recent Judicial and Academic Misconceptions." *Administrative Law Review* 41: 1-32.

_____. 1991. "The Constitutional Limits on Ratemaking: A Reply to Frank Darr." *Energy Law Journal* 12: 111-116.

Posner, Richard A. 1971. "Taxation by Regulation." *Bell Journal of Economics* 2 (Spring): 22-50.

_____. 1974. "Theories of Economic Regulation." *Bell Journal of Economics* 5 (Autumn): 335-358.

Read, James A., Jr. 1989. "Rates of Return that Include New Gas Industry Risks." *Natural Gas* (November): 25-30.

Schuller, Frank C. 1987. "The Roles of Differentiation and Regulation." In *Drawing the Line on Natural Gas Regulation*, edited by Joseph P. Kalt and Frank C. Schuller. New York: Quorum Books.

Shoneman, Charles H. and Gerard R. McConnell. 1986. "FERC Order No. 451: Freedom (Almost) for Old Gas." *Energy Law Journal* 7 (Summer): 299-332.

Soloman, Caleb and Dianna Solis. 1989. "As Northeast's Need for Energy Grows, Gas Becomes a Natural." *The Wall Street Journal*, February 7.

Soloman, Caleb and James Tanner. 1989. "National Gas Prices to Rise in Winter, All Agree, and Some Long-Term Factors Bode Well for Fuel." *The Wall Street Journal*, September 18.

Stalon, Charles G. 1987. "New Challenges to State Regulation of Natural Gas in the Context of FERC Rulemaking." In *Public Utility Regulation in an Environment of Change,* edited by Patrick C. Mann and Harry M. Trebing, pp. 3-13. Lansing, Mich.: Michigan State Universities, MSU Public Utilities Papers, Institute of Public Utilities.

_____. 1990. "The FERC versus the Courts: Sideshow or Main Event in the Structural Transformation of the Natural Gas Industry." With Putnam, Hayes & Bartlett, Inc. Presented at Symposium Assessing the Competitiveness of the Natural Gas Marketplace, sponsored by The Center for Regulatory Studies, Chicago, Illinois, October 9.

Stalon, Charles G. and Reinier H. J. H. Lock. 1990. "State-Federal Relations in the Economic Regulation of Energy." *Yale Journal of Regulation* 7 (Summer): 480-481.

Standard & Poor's. 1991. Industry Surveys. "Utilities -- Electric: Basic Analyses." August 8.

Stigler, George J. 1971. "The Theory of Economic Regulation." *Bell Journal of Economics* 2 (Spring): 3-21.

Train, Kenneth E. 1991. *Optimal Regulation.* Cambridge, Mass.: The MIT Press.

Tye, William B. 1989. "Prudent Investment in Large Complex Projects: The Case of the TransAlaska Pipeline System." *Transportation Practitioners' Journal* 57: 17.

Tye, William B. and A. Lawrence Kolbe. 1992. "Optimal Time Structures for Rates in Regulated Industries." *Transportation Practitioners' Journal* 59 (Winter): 176-196.

U.S. Energy Information Administration. 1989. *Statistics of Interstate Natural Gas Pipeline Companies.* DOE/EIA-0145(89). Washington, D.C.: U.S. Department of Energy.

_____. 1991. *National Energy Strategy.* 1st edition 1991-1992. Springfield, Va.: National Technical Information Service.

Value Line Investment Survey. 1989. January 6.

The Wall Street Journal. 1989a. "Bonds with Resettable Interest Rates Face Major Test with Western Union Note Issue." May 2, p. C2.

_____. 1989b. "Bond Price Softness May Persist as Investors Focus on Inflation." June 19, p. C17.

_____. 1990. "Junk Bond Yields Go Through the Roof." October 11, p. C1.

_____. 1991a. "Panhandle Eastern Plans Write-Down of $280 Million." January 22.

_____. 1991b. "Columbia Gas, Subsidiary File for Chapter 11: Company Seeking to Quit Long-Term Contracts That Lock in High Prices." August 1, p. A3.

_____. 1991c. "Pipe Dream: Glut and a Poor Image Dash the Grand Hopes Held for Natural Gas." August 2, p. A1.

Warzynski, Richard E. 1990. "Strict Scrutiny of FERC Decisions by United States Courts of Appeals." *Energy Law Journal* 11: 269-284.

Webster's New Collegiate Dictionary. 1976. Springfield, Mass.: G. & C. Merriam Co.

Wellisz, Stanislaw H. 1963. "Regulation of Natural Gas Pipeline Companies: An Economic Analysis." *Journal of Political Economy* 71: 30-43.

Whittaker, Win. 1991. "The Discounted Cash Flow Methodology: Its Use in Estimating a Utility's Cost of Equity." *Energy Law Journal* 12: 265-290.

Williams, Stephen F. 1991. "Fixing the Rate of Return After *Duquesne.*" *Yale Journal on Regulation* 8 (Winter): 159-163.

Index